Protein Dynamics, Function, and Design

NATO ASI Series

Advanced Science Institutes Series

A series presenting the results of activities sponsored by the NATO Science Committee, which aims at the dissemination of advanced scientific and technological knowledge, with a view to strengthening links between scientific communities.

The series is published by an international board of publishers in conjunction with the NATO Scientific Affairs Division

A	**Life Sciences**	Plenum Publishing Corporation
B	**Physics**	New York and London
C	**Mathematical**	Kluwer Academic Publishers
	and Physical Sciences	Dordrecht, Boston, and London
D	**Behavioral and Social Sciences**	
E	**Applied Sciences**	
F	**Computer and Systems Sciences**	Springer-Verlag
G	**Ecological Sciences**	Berlin, Heidelberg, New York, London,
H	**Cell Biology**	Paris, Tokyo, Hong Kong, and Barcelona
I	**Global Environmental Change**	

PARTNERSHIP SUB-SERIES

1. **Disarmament Technologies**	Kluwer Academic Publishers
2. **Environment**	Springer-Verlag
3. **High Technology**	Kluwer Academic Publishers
4. **Science and Technology Policy**	Kluwer Academic Publishers
5. **Computer Networking**	Kluwer Academic Publishers

The Partnership Sub-Series incorporates activities undertaken in collaboration with NATO's Cooperation Partners, the countries of the CIS and Central and Eastern Europe, in Priority Areas of concern to those countries.

Recent Volumes in this Series:

Series A: Life Sciences

Protein Dynamics, Function, and Design

Edited by

Oleg Jardetzky

Stanford University
Stanford, California

and

Jean-François Lefèvre

Université Louis Pasteur
Illkirch Graffenstaden, France

Assistant Editor

Robin E. Holbrook

Stanford University
Stanford, California

Plenum Press
New York and London
Published in cooperation with NATO Scientific Affairs Division

Proceedings of a NATO Advanced Study Institute and
International School of Structural Biology and Magnetic Resonance, 3rd Course on
Protein Dynamics, Function, and Design
held April 16 – 28, 1997,
in Erice, Italy

NATO-PCO-DATA BASE

The electronic index to the NATO ASI Series provides full bibliographical references (with keywords and/or abstracts) to about 50,000 contributions from international scientists published in all sections of the NATO ASI Series. Access to the NATO-PCO-DATA BASE is possible via a CD-ROM "NATO Science and Technology Disk" with user-friendly retrieval software in English, French, and German (©WTV GmbH and DATAWARE Technologies, Inc. 1989). The CD-ROM contains the AGARD Aerospace Database.

The CD-ROM can be ordered through any member of the Board of Publishers or through NATO-PCO, Overijse, Belgium.

Library of Congress Cataloging-in-Publication Data

QP
551
.P6958177
1998

Protein dynamics, function, and design / edited by Oleg Jardetzky and
   Jean-François Lefévre ; assist. editor, Robin E. Holbrook.
        p.   cm. -- (NATO ASI series. Series A, Life sciences ; v.
   301)
     "Proceedings of NATO Advanced Study Institute and International
   School of Structural Biology and Magnetic Resonance, 3rd Course on
   Protein Dynamics, Function, and Design, held April 16-28, 1997, in
   Erice, Italy"--T.p. verso.
     Includes bibliographical references and index.
     ISBN 0-306-45939-6
     1. Proteins--Structure-activity relationships--Congresses.
   I. Jardetzky, Oleg. II. Lefévre, Jean-François. III. Holbrook,
   Robin. IV. NATO Advanced Study Institute and International School
   of Structural Biology and Magnetic Resonance: 3rd Course on Protein
   Dynamics, Function, and Design (1997 : Erice, Italy) V. Series.
   QP551.P6958177   1998
   572'.6--dc21
                                                98-41432
                                                CIP

ISBN 0-306-45939-6

© 1998 Plenum Press, New York
A Division of Plenum Publishing Corporation
233 Spring Street, New York, N.Y. 10013

http://www.plenum.com

10 9 8 7 6 5 4 3 2 1

PREFACE

This volume is a collection of articles from the proceedings of the International School of Structural Biology and Magnetic Resonance 3rd Course: Protein Dynamics, Function, and Design. This NATO Advance Study Institute was held in Erice at the Ettore Majorana Centre for Scientific Culture on April 16–28, 1997. The aim of the Institute was to bring together experts applying different physical methods to problems of macromolecular dynamics—notably x-ray diffraction, NMR and other forms of spectroscopy, and molecular dynamics simulations. Emphasis was placed on those systems and types of problems—such as mechanisms of allosteric control, signal transmission, induced fit to different ligands with its implications for drug design, and the effects of dynamics on structure determination—where a correlation of findings obtained by different methods could shed the most light on the mechanisms involved and stimulate the search for new approaches. The individual articles represent the state of the art in each of the areas covered and provide a guide to the original literature in this rapidly developing field.

CONTENTS

DETERMINING STRUCTURES OF PROTEIN/DNA COMPLEXES BY NMR

Angela M. Gronenborn and G. Marius Clore

Laboratory of Chemical Physics
National Institute of Diabetes and Digestive and Kidney Diseases
National Institutes of Health
Bethesda, Maryland 20892-0520

INTRODUCTION

It is axiomatic that a detailed understanding of the function of macromolecules necessitates knowledge of their three-dimensional structures. At the present time, the two main techniques that can provide a compete description of the structures of macromolecules at the atomic level are X-ray crystallography in the solid state (single crystals) and nuclear magnetic resonance spectroscopy in solution. Both techniques have different advantages and drawbacks and taken together yield highly complementary results.

The size of macromolecular structures that can be solved by NMR has been dramatically increased over the last decades. The development of a wide range of 2D NMR experiments in the early 1980's culminated in the determination of the structures of a number of small proteins (Wüthrich, K., 1986 ; Clore, G.M. and Gronenborn, A.M., 1987). Under exceptional circumstances, 2D NMR techniques can be applied successfully to the structure determination of proteins up to ~100 residues (Dyson et al., 1990; Forman-Kay et al., 1991). Beyond 100 residues, however, 2D NMR methods tend to fail, principally due to spectral complexity which cannot be resolved in two dimensions. In the late 1980's and early 1990's, a series of major advances took place in which the spectral resolution was increased by extending the dimensionality to three- and four-dimensions (Clore, G.M. and Gronenborn, A.M., 1991a). In addition, by combining such multidimensional experiments with heteronuclear NMR, problems associated with large linewidths can be circumvented by making use of heteronuclear couplings that are large relative to the linewidths. Concomitant with the spectroscopic advances, significant improvements have taken place in the accuracy with which macromolecular structures can be determined. Key improvements in refinement techniques include direct refinement against accurate coupling constants (Garrett et al., 1994) and ^{13}C and ^{1}H shifts (Kuszewski et al., 1995a; 1995b), as well as the use of conformational database potentials (Kuszewski et al., 1996; 1997). More re-

Protein Dynamics, Function, and Design, edited by Jardetzky *et al.*
Plenum Press, New York, 1998

cently, new methods have been developed to obtain structural restraints that characterize long range order *a priori* (Tjandra et al., 1997a; 1997b). These include making use of the dependence of heteronuclear relaxation on the rotational diffusion anisotropy of non-spherical molecules and of the field dependence of one-bond heteronuclear couplings arising from magnetic susceptibility anisotropy.

Structures of protein-DNA complexes determined by multidimensional NMR include those of DNA binding domains from the transcription factor GATA-1 (Omichinski et al., 1993), the *lac* repressor headpiece (Chuprina et al., 1993), the antennapedia homeodomain (Billeter et al., 1993), the protooncogene c-myb (Ogata et al., 1994), the *trp* repressor (Zhang et al., 1994), the human testis determining factor SRY (Werner et al., 1995), the lymphoid enhancer binding factor LEF-1 (Love et al., 1995), the chromatin remodeling factor GAGA (Omichinski et al., 1997), and the transcriptional coactivator HMG-I(Y) (Huth et al., 1997).

SAMPLE REQUIREMENTS FOR NMR SPECTROSCOPY

In the study of macromolecules or complexes thereof, concentrations of about 1 mM are typically employed in a sample volume of 0.3–0.5 ml. A key requirement is that the complex under study should be soluble, should not aggregate, and should be stable for many weeks at room temperature. For ^{1}H homonuclear work, it is also important to ensure that the buffer employed does not contain any protons. In general, two samples are required, one in D_2O for the observation of non-exchangeable protons only, and the other in 95% H_2O/5% D_2O to permit the observation of exchangeable protons.

While it is possible to use ^{1}H homonuclear methods to solve structures of proteins up to about 100 residues in certain very favourable cases (Dyson et al., 1990; Forman-Kay et al., 1991), it is generally the case that extensive resonance overlap makes this task very time consuming and complex. Hence, providing a protein can be overexpressed in a bacterial system, it is now desirable, even for proteins as small as 30 residues, to make use of the full panoply of multidimensional heteronuclear NMR experiments (Clore and Gronenborn, 1991b; Bax and Grzesiek, 1993). This is clearly a necessary requirement for structure determinations of complexes. Uniform ^{15}N and ^{13}C labeling is achieved by growing the bacteria on minimal medium containing ^{15}NH$_4$Cl and ^{13}C$_6$-glucose as the sole nitrogen and carbon sources, respectively. In general, the following protein samples are required: ^{15}N-labeled sample in 95% H_2O/5% D_2O, ^{15}N/^{13}C-labeled sample in 95% H_2O/5% D_2O, and ^{15}N/^{13}C-labeled sample in D_2O. For very large systems, deuteration and specific labeling is advantageous (Grzesiek et al., 1995; Venters et al., 1995). For example, a particularly useful strategy for aromatic residues is one of reverse labeling in which the aromatics are at natural isotopic abundance, while the other residues are ^{13}C/^{15}N labeled (Vuister et al., 1994). This can be achieved by adding the aromatic amino acids at natural isotopic abundance to the minimal medium in addition to ^{15}NH$_4$Cl and ^{13}C$_6$-glucose. A similar approach can also be used for various aliphatic amino acids. Deuteration of non-exchangeable protons is achieved by growing the bacteria in D_2O as opposed to H_2O medium.

NMR SPECTROSCOPY ASSIGNMENT

We generally use sequential assignment strategy based solely on well defined heteronuclear scalar couplings along the polypeptide chain (Clore and Gronenborn, 1991b;

Table 1. Summary of correlations observed in the 3D double and triple resonance experiments used for sequential and side chain assignments

Experiment	Correlation	J Coupling[a]
^{15}N-edited HOHAHA	C$^{\alpha}$H(i)-^{15}N(i)-NH(i)	$^3J_{HN\alpha}$
	C$^{\beta}$H(i)-^{15}N(i)-NH(i)	$^3J_{HN\alpha}$ and $^3J_{\alpha\beta}$
HNHA	C$^{\alpha}$H(i)-^{15}N(i)-NH(i)	$^3J_{HN\alpha}$
H(CA)NH	C$^{\alpha}$H(i)-^{15}N(i)-NH(i)	$^1J_{NC\alpha}$
	C$^{\alpha}$H(i-1)-^{15}N(i)-NH(i)	$^2J_{NC\alpha}$
HNCA	^{13}C$^{\alpha}$(i)-^{15}N(i)-NH(i)	$^1J_{NC\alpha}$
	^{13}C$^{\alpha}$(i-1)-^{15}N(i)-NH(i)	$^2J_{NC\alpha}$
HN(CO)CA	^{13}C$^{\alpha}$(i-1)-^{15}N(i)-NH(i)	$^1J_{NCO}$ and $^1J_{C\alpha CO}$
HNCO	^{13}CO(i-1)-^{15}N(i)-NH(i)	$^1J_{NCO}$
HCACO	C$^{\alpha}$H(i)-^{13}C$^{\alpha}$(i)-^{13}CO(i)	$^1J_{C\alpha CO}$
HCA(CO)N	C$^{\alpha}$H(i)-^{13}C$^{\alpha}$(i)-^{15}N(i+1)	$^1J_{C\alpha CO}$ and $^1J_{NCO}$
CBCA(CO)NH	^{13}C$^{\beta}$(i-1)/^{13}C$^{\alpha}$(i-1)-^{15}N(i)-NH(i)	$^1J_{C\alpha CO}$
CBCANH	^{13}C$^{\beta}$(i)/^{13}C$^{\alpha}$(i)-^{15}N(i)-NH(i)	$^1J_{NC\alpha}$ and $^1J_{CC}$
	^{13}C$^{\beta}$(i-1)/^{13}C$^{\alpha}$(i-1)-^{15}N(i)-NH(i)	$^2J_{NC\alpha}$ and $^1J_{CC}$
HBHA(CO)NH	C$^{\beta}$H(i-1)/C$^{\alpha}$H(i-1) -^{15}N(i)-NH(i)	$^1J_{C\alpha CO}$
HBHA(CBCA)NH	C$^{\beta}$H(i)/C$^{\alpha}$H(i) -^{15}N(i)-NH(i)	$^1J_{NC\alpha}$ and $^1J_{CC}$
	C$^{\beta}$H(i-1)/C$^{\alpha}$H(i-1) -^{15}N(i)-NH(i)	$^2J_{NC\alpha}$ and $^1J_{CC}$
C(CO)NH	^{13}C^j(i-1)-^{15}N(i)-NH(i)	$^1J_{C\alpha CO}$
H(CCO)NH	H^j(i-1) -^{15}N(i)-NH(i)	$^1J_{C\alpha CO}$
HCCH-COSY	H^j-^{13}C^j-^{13}C$^{j\pm1}$-H$^{j\pm1}$	$^1J_{CC}$
HCCH-TOCSY	H^j-^{13}C^j....^{13}C$^{j\pm n}$-H$^{j\pm n}$	$^1J_{CC}$

[a]In addition to the couplings indicated, all the experiments make use of the $^1J_{CH}$ (~140 Hz) and/or $^1J_{NH}$ (~95 Hz) couplings. The values of the couplings employed are as follows: $^3J_{HN\alpha}$ ~3–10 Hz, $^1J_{CC}$ ~35 Hz, $^1J_{C\alpha CO}$ ~55 Hz, $^1J_{NCO}$ ~15 Hz, $^1J_{NC\alpha}$ ~11 Hz, $^2J_{NC\alpha}$ ~7 Hz.

Bax and Grzesiek, 1993). The double and triple resonance experiments that we currently use together with the correlations that they demonstrate, are summarized in Table 1. With the advent of pulsed field gradients to either eliminate undesired coherence transfer pathways (Bax and Pochapsky, 1992) or to select particular coherence pathways coupled with sensitivity enhancement (Kay et al., 1992), it is now possible to employ only two to four step phase-cycles without any loss in sensitivity (other than that due to the reduction in measurement time) such that each 3D experiment can be recorded in as little as 7 hours. In most cases, however, signal-to-noise requirements necessitate 1–3 days measuring time depending on the experiment. For systems greater than ~25 kDa, the assignment of the backbone and sidechain carbons of the protein is facilitated by making use of a sample in which the non-exchangeable protons are deuterated, thereby dramatically reducing the linewidths.

Providing the DNA yields a relatively simple spectrum that can be assigned by 2D methods, the most convenient strategy for dealing with protein-DNA complexes involves one in which the protein is labeled with ^{15}N and ^{13}C and the DNA is unlabeled (i.e. at natural isotopic abundance). It is then possible to use a combination of heteronuclear filtering and editing (Otting & Wüthrich, 1990; Ikura & Bax, 1992) to design experiments in which correlations involving only protein resonances, only DNA resonances, or only through-space interactions between ligand and DNA are observed. These experiments are summarized in Tables 2 and 3, and have been successfully employed in a number of laboratories for a range of systems in addition to protein/DNA complexes. Sequential assignments of the exchangeable and non-exchangeable protons of the DNA in a complex are obtained by standard procedures using a 2D NOE spectrum in water with a 1–1 semi-selective excita-

Table 2. Summary of heteronuclear-filtered and -separated NOE experiments used to study protein-DNA complexes comprising a uniformly ^{15}N/^{13}C labeled protein and unlabeled DNA

Type of contact	Connectivity
A. Intra- and intermolecular contacts	
3D ^{15}N-edited NOE in H$_2$O	H(j)-^{15}N(j)----------H(i)
3D ^{13}C-edited NOE in D$_2$O	H(j)-^{13}C(j)----------H(i)
B. Intramolecular protein contacts	
4D ^{13}C/^{13}C-edited NOE in D$_2$O	H(j)-^{13}C(j)----------H(i)-^{13}C(i)
4D ^{15}N/^{13}C-edited NOE in H$_2$O	H(j)-^{15}N(j)----------H(i)-^{13}C(i)
3D ^{15}N/^{15}N-edited NOE in H$_2$O	H(j)-^{15}N(j)----------H(i)-^{15}N(i)
C. Intramolecular DNA contacts	
2D ^{12}C,^{14}N(F$_1$)/ ^{12}C,^{14}N(F$_2$) filtered NOE in H$_2$O[a]	H(j)-^{12}C(j)-------H(i)-^{12}C(i)
	H(j)-^{14}N(j)-------H(i)-^{12}C(i)
	H(j)-^{12}C(j)-------H(i)-^{14}N(i)
	H(j)-^{14}N(j)-------H(i)-^{14}N(i)
2D ^{12}C(F$_1$)/^{12}C(F$_2$) filtered NOE in D$_2$O[a]	H(j)-^{12}C(j)-------H(i)-^{12}C(i)
D. Intermolecular protein-DNA contacts	
3D ^{15}N-edited(F$_1$)/^{14}N,^{12}C(F$_3$) filtered NOE in H$_2$O	H(j)-^{15}N(j)-------H(i)-^{12}C(i)
	H(j)-^{15}N(j)-------H(i)-^{14}N(i)
3D ^{13}C-edited(F$_1$)/^{12}C(F$_3$) filtered NOE in D$_2$O	H(j)-^{13}C(j)-------H(i)-^{12}C(i)

[a]Similar heteronuclear-filtered 2D correlation and Hartmann-Hahn spectra can also be recorded to assign the spin systems of the DNA.

tion pulse to assign the exchangeable and A(H2) protons, and 2D ^{12}C-filtered NOE and HOHAHA spectra in D$_2$O to observe correlations involving only protons attached to ^{12}C, suppressing the signals of ^{13}C attached protons from the protein.

ASSIGNMENT OF NOES WITHIN THE PROTEIN, THE DNA AND BETWEEN PROTEIN AND DNA

While the panoply of 3D heteronuclear experiments is sufficient for the purposes of spectral assignment, yet further increases in resolution are required for the reliable identification of NOE through-space interactions. This can be achieved by extending the dimensionality still further to four dimensions (Kay et al., 1990; Clore et al., 1991). In this manner, each ^{1}H-^{1}H NOE interaction is specified by four chemical shift coordinates, the two protons giving rise to the NOE and the heavy atoms to which they are attached.

Because the number of NOE interactions present in each 2D plane of a 4D ^{13}C/^{15}N or ^{13}C/^{13}C-edited NOESY spectrum is so small, the inherent resolution in a 4D spectrum is

Table 3. Additional heteronuclear-filtered and -separated NOE experiments that can be used to selectively detect intermolecular NOEs in protein-DNA complexes employing labeled DNA

Labeling	Connectivity
U-^{15}N/^{2}H + U-^{13}C/^{1}H	
3D ^{13}C-filtered/^{15}N-separated	H(j)-^{13}C(j)----------H(i)-^{15}N(i)
3D ^{13}C-separated/^{15}N-filtered	H(j)-^{13}C(j)----------H(i)-^{15}N(i)
4D ^{13}C-separated/^{15}N-separated	H(j)-^{13}C(j)----------H(i)-^{15}N(i)

extremely high, despite the low level of digitization. Further, it can be calculated that 4D spectra with virtual lack of resonance overlap and good sensitivity can potentially be obtained on proteins with as many as 400 residues. Thus, once complete ^{1}H, ^{15}N and ^{13}C assignments are obtained, analysis of 4D $^{15}N/^{13}C$ and $^{13}C/^{13}C$-separated NOE spectra should permit the assignment of many NOE interactions in a relatively straightforward manner (Clore and Gronenborn, 1991b).

NOEs involving only the DNA are detected using the same 2D NOE and ^{12}C-filtered NOE spectra, recorded in H_2O and D_2O, respectively, employed for assignment purposes. NOEs involving protons of the protein alone are obtained from 3D ^{15}N- and ^{13}C-separated NOESY spectra recorded in H_2O and D_2O, respectively. Finally NOEs specifically between protein and DNA protons are identified in 3D ^{12}C-filtered/^{13}C-separated and $^{12}C,^{14}N$-filtered/^{15}N-separated NOESY spectra, recorded in D_2O and H_2O, respectively.

COUPLING CONSTANTS AND TORSION ANGLES

Torsion angle restraints can be derived from coupling constant data, since there exist simple geometric relationships between three-bond couplings and torsion angles. In simple systems, the coupling constant can be measured directly from the in-phase or anti-phase splitting of a particular resonance in the 1D or 2D spectrum. For larger systems where the linewidths exceed the coupling, it becomes difficult to extract accurate couplings in this manner. An alternative approach involves the use of ECOSY experiments to generate reduced cross-peak multiplets (Griesinger at al., 1986). While this permits accurate couplings to be obtained, the sensitivity of ECOSY experiments is generally quite low. Furthermore, in multidimensional experiments its utility is restricted by the fact that the couplings have to be measured in the indirectly detected dimensions, and hence are influenced by limited digital resolution. More recently, a series of highly sensitive quantitative J correlation experiments have been developed which circumvent these problems (Bax et al., 1994). These experiments quantitate the loss in magnetization when dephasing caused by coupling is active versus inactive. In some J quantitative correlation experiments, the coupling is obtained from the ratio of cross peak to diagonal peak intensities. In others, it is obtained by the ratio of the cross peaks obtained in two separate experiments (with the coupling active and inactive), recorded in an interleaved manner.

A summary of the heteronuclear quantitative J correlation experiments that we currently employ is provided in Table 4.

EXAMPLES

The GATA-1/DNA Complex

The erythroid specific transcription factor GATA-1 is responsible for the regulation of transcription of erythroid-expressed genes and is an essential component required for the generation of the eyrthroid lineage (Orkin, 1992). GATA-1 binds specifically as a monomer to the asymmetric consensus target sequence (T/A)GATA(A/G) found in the cis-regulatory elements of all globin genes and most other erythroid specific genes that have been examined (Evans & Felsenfeld, 1989). GATA-1 was the first member of a family of proteins, which contain metal binding regions of the form Cys-X-X-Cys-$(X)_{17}$-Cys-X-X-Cys. We determined the solution structure of the specific complex of a 66 residue frag-

Table 4. Experiments for determining three-bond coupling
constants by quantitative J correlation spectroscopy[a]

Experiment	Three-bond coupling	Torsion angle
3D HNHA	$^3J_{HN\alpha}$	ϕ
3D (HN)CO(CO)NH	$^3J_{COCO}$	ϕ
2D ^{13}C-$\{^{15}$N$\}$-spin-echo difference CT-HSQC	$^3J_{C\gamma N}$	χ_1 of Thr and Val
2D ^{13}C-$\{^{13}$CO$\}$-spin-echo difference CT-HSQC	$^3J_{C\gamma CO}$	χ_1 of Thr and Val
2D ^{13}CO-$\{^{13}$Cγ(aro)$\}$ spin-echo difference ^{1}H-^{15}N HSQC	$^3J_{C\gamma(aromatic)CO}$	χ_1 of aromatics
2D ^{15}N-$\{^{13}$Cγ(aro)$\}$ spin-echo difference ^{1}H-^{15}N HSQC	$^3J_{C\gamma(aromatic)N}$	χ_1 of aromatics
2D ^{15}N-$\{^{13}$C$\gamma\}$ spin-echo difference ^{1}H-^{15}N HSQC	$^3J_{C\gamma(aliphatic)N}$	χ_1 of aliphatics
3D HN(CO)C	$^3J_{C\gamma(aliphatic)CO}$	χ_1 of aliphatics
3D HN(CO)HB	$^3J_{COH\beta}$	χ_1
3D HNHB	$^3J_{NH\beta}$	χ_1
3D HACAHB	$^3J_{\alpha\beta}$	χ_1
2D or 3D ^{1}H-detected long-range C-C COSY	$^3J_{CC}$	χ_2 of Leu and Ile χ_3 of Met
3D ^{1}H-detected [^{13}C-^{1}H] long-range COSY	$^3J_{CH}$	χ_2 of Leu and Ile χ_3 of Met

[a] Details of most experiments are provided in the review by Bax et al. (1994). The 2D ^{13}CO-$\{^{13}$Cγ(aro)$\}$, ^{15}N-$\{^{13}$Cγ(aro)$\}$, and ^{15}N-$\{^{13}$C$\gamma\}$ spin-echo difference ^{1}H-^{15}N HSQC experiments are described in Grzesiek and Bax (1997) and Hu and Bax (1997).

ment (residues 158–223) comprising the DNA binding domain of chicken GATA-1 (cGATA-1) with a 16 base pair oligonucleotide containing the target sequence AGATAA by NMR spectroscopy (Omichinski et al., 1993).

The protein can be divided into two modules: the protein core which consists of residues 2–51 and contains the zinc coordination site, and an extended C-terminal tail (residues 52–59). Part of the core of the cGATA-1 DNA binding domain is structurally similar to that of the N-terminal zinc containing module of the DNA binding domain of the glucocorticoid receptor (Luisi et al., 1991).

The overall topology and structural organization of the complex is shown in Fig. 3 The conformation of the oligonucleotide is B-type. The helix and the loop connecting strands β2 and β3 (which is located directly beneath the helix) are located in the major groove, while the C-terminal tail wraps around the DNA and lies in the minor groove, directly opposite the helix.

The mode of specific DNA binding protein that is revealed in this structure is distinct from that observed for the other three classes of zinc containing DNA binding domains whose structures have previously been solved (Pavletich & Pabo, 1991, 1993; Luisi et al., 1991; Mamorstein et al., 1992; Schwabe et al., 1993; Fairall et al., 1993). Features specific to the complex with the DNA binding domain of cGATA-1 include the relatively small size of the DNA target site (8 base pairs of which only a contiguous stretch of 6 is involved in specific contacts), the monomeric nature of the complex in which only a *single* zinc binding module is required for specific binding, the predominance of hydrophobic interactions involved in specific base contacts in the major groove, the presence of a basic

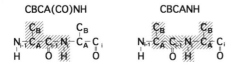

Figure 1. Schematic illustration of correlations along the polypeptide chain observed in the CBCA(CO)NH and CBCANH experiments.

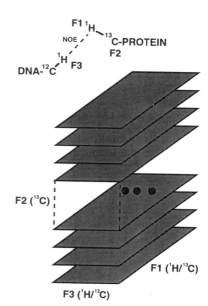

Figure 2. Schematic diagram illustrating the principles of a 3D heteronuclear filtered and edited experiment. In the 3D spectrum, the NOE cross-peaks between ^{12}C attached protons and ^{13}C attached protons appear in a plane, edited according to the shift of the heavy atom (^{13}C) attached to the originating proton.

C-terminal tail which interacts with the DNA in the minor groove and constitutes a key component of specificity, and finally the pincer-like nature of the complex in which the core and tail subdomains are opposed and surround the DNA just like a hand gripping a rope. The structure of the cGATA-1 DNA binding domain reveals a modular design. The fold of residues 3–39 is similar to that of the N-terminal zinc binding module of the DNA binding domain of the glucocorticoid receptor, although, with the exception of the four Cys residues that coordinate zinc, there is no significant sequence identity between these regions of the two proteins. Residues 40–66 are part of a separate structural motif. In this regard it is interesting to note that, in addition to both zinc binding modules being encoded on separate exons in the cGATA-1 gene, the next intron/exon boundary lies between amino acids 39 and 40 of the DNA binding domain, thereby separating the C-terminal zinc binding domain from the basic tail (Hannon et al., 1991).

The GAGA/DNA Complex

The GAGA factor of *Drosophila melanogaster* is a TFIIIA-like zinc finger protein, originally identified on the basis of its ability to bind to $(GA)_n$ rich sites in promoters, which acts as an anti-repressor by helping to disrupt nucleosomes associated with gene regulatory sequences. The protein is 519 residues in length and comprises three domains: an N-terminal POZ/PTB protein interaction domain, a central DNA binding domain, and a polyglutamine-rich carboxy-terminus (Granok et al., 1995). The minimal DNA binding domain (GAGA-DBD) binds specifically to DNA derived from the h3/h4 promoter, containing the sequence GAGAGAG (Pedone et al., 1996). The GAGA-DBD consists of a single classical Cys$_2$-His$_2$ zinc finger (Klug and Schwabe, 1995; Berg and Shi, 1996) preeceded by two highly basic regions (BR1 and BR2). Although there is also a basic region C-terminal to the zinc finger domain present, it is not important for DNA binding. We determined the three-dimensional structure of a complex between the GAGA-DBD and an oligonucleotide containing its GA-GAG consensus binding site (Omichinski et al., 1997). The overall structural organization of the complex is illustrated in Fig. 4. The zinc finger core is centered around a tetrahedrally co-ordinated zinc atom, comprises a $\beta\beta\alpha$ motif and is very similar to that of other classical zinc fingers. It binds in the major groove and recognizes the first three GAG bases of the consensus in a manner very similar to that seen in other zinc finger/DNA complexes. Since the $H^{\delta 1}$ proton of the imidazole ring of a Zn coordinating His is hydrogen-bonded to a backbone phosphate its resonance is visible in the ^1H NMR spectrum at ~14.5 ppm, and this interaction is probably responsible for the ~0.7 ppm downfield shift in the H5" resonance of C3. In contrast to other classical zinc fingers which comprise repeated units with a minimum of two for high affinity binding, the GAGA-DBD makes use of only a single finger complemented by the two N-terminal basic regions BR1 and BR2. BR2 forms a helix that interacts in the major groove recognizing the last G of the consensus, while BR1 wraps around the DNA in the minor groove and recognizes the A in the fourth position of the consensus. The DNA in the complex is essentially B-type with a minor smooth bend of ~10°. The general mode of DNA binding (as opposed to the specifics of sidechain-base interactions and recognition) observed for the GAGA-DBD is reminiscent of that seen for the DBD of the GATA-1 transcription factor (Omichinski et al., 1993). The GATA-1-DBD also clamps the DNA with a helix and loop interacting in the major groove, and a basic C-terminal tail wrapping around the minor groove. The size of the DNA binding site is similar, and, just as for the GAGA-DBD, the major groove interactions involve a zinc binding domain. The zinc binding module of GATA-1, however, is not a classical TFIIIA-like zinc finger but is structurally related to the amino-terminal zinc module of the glucocorticoid receptor (Luisi et al., 1991). Thus, both the GAGA-DBD and the GATA-1-DBD employ two structural motifs to interact with the DNA.

The structure of the GAGA-DBD complex shows how a classical zinc finger complemented by an N-terminal extension comprising a basic helix and tail can recognize DNA in a sequence specific manner, making base specific contacts with every base in the pentanucleotide consensus sequence GAGAG. In addition, the presence of both major and minor groove contacts in the complex immediately suggests special constraints on interactions of the GAGA factor and DNA targets within nucleosomes.

The Structure of the Human SRY-DNA Complex

Male sex determination in mammals is directed by the genetic information encoded on the Y chromosome (Goodfellow and Lovell-Badge,1993; Haqq et al., 1994). A key component is the protein encoded by the human testis determining gene, SRY (Sex determining Region Y). Mutations in this gene are responsible for 15% of male to female sex reversal, that is 46X,Y females (Goodfellow and Lovell-Badge, 1993; Gustafson and Donahoe, 1994). The SRY protein is a transcriptional activator containing 203 residues

Figure 3. Schematic ribbon drawings illustrating the interactions of cGATA-1 with DNA. The left hand side panels show views into the major groove while the right hand side ones show views into the minor groove. Side chain interactions between cGATA-1 and the DNA are shown in the bottom two views. The protein backbone is shown in green and the protein side chains in yellow; the color code for the DNA bases is as follows: red for A, lilac for T, dark blue for G and light blue for C. (From Omichinski et al., 1993).

Figure 4. Four views illustrating the interaction of the GAGA-DBD with DNA. The backbone of the protein is depicted as a green tube, side chains making base specific contacts are in yellow, the histidine and cysteine side chains coordinating the zinc (blue ball) in magenta, GC base pairs in red, and AT base pairs in blue. In (**B**) the DNA is depicted as a molecular surface with the bases colored in blue and the sugar-phosphate backbone in white. The path of the long axis of the DNA helix is shown in yellow in (**C**). The structure shown is the restrained regularized mean structure. (From Omichinski et al., 1997).

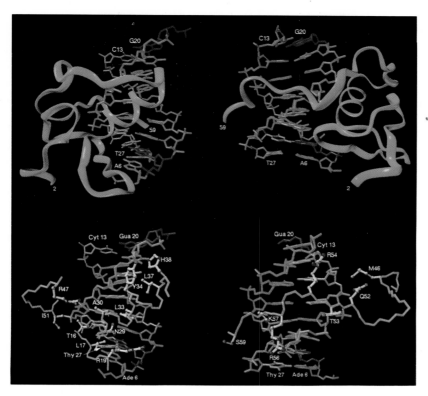

Figure 3.

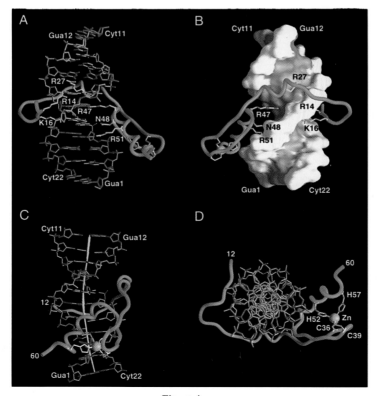

Figure 4.

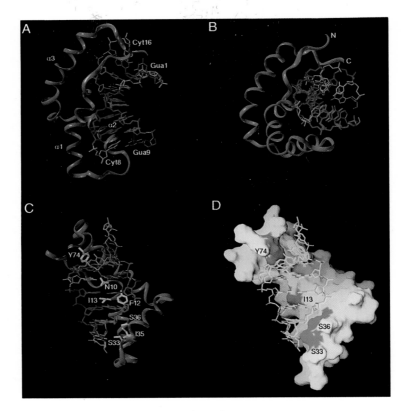

Figure 5.

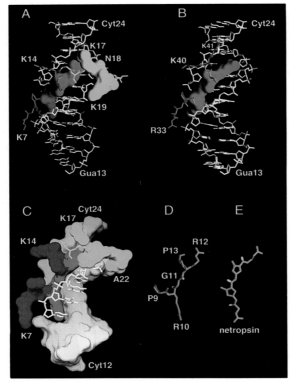

Figure 6.

which comprise three domains: an N-terminal domain, a central 77 residue DNA binding domain consisting of a single HMG (high mobility group) box, and a C-terminal domain (Sinclair et al., 1990). The HMG box is an approximately 80 residue domain which mediates the minor groove DNA binding of a large family of eukaryotic proteins known as the HMG-1/2 family (Laudet et al., 1993). We have determined the solution structure of a specific complex of the DNA binding domain of hSRY (hereafter referred to as hSRY-HMG) with a DNA octamer comprising a specific target site in the MIS promoter (Werner et al., 1995). Several pertinent features of this complex are illustrated in Fig. 5.

The overall topology of hSRY-HMG is best described as a twisted letter L or boomerang with the long and short arms approximately 28 Å and 22 Å in length, respectively. The structure is composed of irregular N- and C- terminal strands which lie directly opposite each other, although they are not hydrogen bonded, and three helices. The long arm of the L is formed by helix 3 and the N-terminal strand, while the short arm of the L is formed by helices 1 and 2. The orientation of the long and short arms is maintained by the packing of several side chains located both on helix 1 and helix 3. The top of the long arm is held in place by a hydrophobic cluster between the N- and C-termini.

Upon binding to hSRY-HMG, the DNA undergoes a profound structural change from B-type DNA in the free state to an underwound form that is distinct from either B or A type DNA . The DNA helix is severely underwound and, as a result, the minor groove is shallow and significantly expanded with a width of 9.4±0.6 Å compared to 4.0 Å in B-DNA and 6.2 Å in A-DNA. Concomitantly, the major groove is compressed. The DNA is bent by ~70–80° in the direction of the major groove which is accomplished through large positive local interbasepair roll angles for 6 of the 7 basesteps in conjuction with helical unwinding.

The DNA is located in the concave surface of the L-shaped hSRY-HMG domain and binding occurs exclusively in the minor groove of the DNA. The binding surface is formed by helices 1 and 3, bounded at the bottom by a ridge comprising helix 2 and at the top by a ridge comprising the N- and C-terminal strands. The two principal areas of bending occur between basepairs 5 and 6 as a result of the partial intercalation of an Ile sidechain, and between basepairs 2 and 3, as the DNA is pushed away from the body of the protein by the ridge formed by a Lys and a Tyr. Widening of the minor groove appears to be mediated by five residues that form a T-shaped wedge in direct contact with the central basepairs of the DNA octamer. Specifically, four hydrophobic residues form a hydro-

Figure 5. Four views illustrating the interaction of hSRY-HMG with DNA. The protein is shown as a schematic ribbon drawing in green, and the color coding used for the DNA bases is red for A, lilac for T, dark blue for G and light blue for C. Side chains that contact the DNA bases are depicted in yellow in (C). (D) Same view as in (C) with the molecular surface of the protein shown in grey and the DNA atoms in yellow. The patches of blue on the protein surface indicate the location of the side chains of 4 of the 7 residues that interact with the DNA bases. The protrusion on the surface of the protein associated with the side chain of Ile13, which partially intercalates between base pairs 5 and 6 of the DNA, is clearly visible in (D). (From Werner et al., 1995).

Figure 6. Illustrations of (A) DBD2 and (B) DBD3 of HMG-I(Y) with DNA in which the core Arg-Gly-Arg DNA-binding unit is shown in blue, and the polar network of amino acids that accounts for the higher affinity of DBD2 relative to the DBD3 is shown in yellow. The DNA and the backbone and sidechains of the proline and lysine residues immediately adjacent to the core are displayed as a bond representation in white and red, respectively. (C) View of DBD2 where the DNA and modular components of the type-I motif are color coded: white for the DNA, blue for the Arg-Gly-Arg core, red for amino acids that contact the phosphate backbone similar to those in the type-II motif, and yellow for the DNA-contacting residues unique to the type-I motif. (D) Bond representation of the Arg-Gly-Arg core and adjacent prolines. Carbons are colored in blue, nitrogens in green, oxygens in yellow, and the prolines in red. (E) Bond representation of netropsin with the same color coding as for panel (D). The structures shown are the restrained regularized mean structures.

phobic wedge across basepairs 5 and 6, anchored to basepairs 4 and 5 by electrostatic interactions involving an Asn. The residues constituting the central portion of the wedge (Phe and Ile) and the stem of the T (Asn) bind to the DNA bases, while the residues at the wings of the wedge (Met and Trp) bind to the DNA sugar-phosphate backbone and pry open the minor groove. This T-shaped wedge on the surface of hSRY-HMG can be considered to directly induce and stabilize the helix unwinding of the DNA. In addition, hSRY-HMG is anchored at the two ends of the DNA by additional interactions.

To date kinking the DNA by minor groove intercalation of one or more amino acids has been observed for several proteins involved in transcriptional regulation. Specifically, the high-mobility group (HMG) domain proteins SRY (Werner et al., 1995) and LEF-1 (Love et al., 1995), the TATA-box binding protein TBP (*Arabidopsis thaliana*, aTBP; yeast, yTBP (Kim, Y. et al., 1993; Kim, J.L. et al., 1993) and integration host factor (Rice et al., 1996) utilize an intercalative wedge to pry open a single basestep and distort the DNA. Strikingly, each protein has a completely different topology with different parts of the structures being involved in the interactions with DNA.

The Structure of an HMG-I(Y) Complex

Members of the non-histone chromosomal high mobility group HMG-I(Y) family (Bustin and Reeves, 1996) are required for the assembly of higher order transcription enhancer complexes which are central for transcriptional activation of a number of important genes, the best characterized of which is the interferon-β gene (IFN-β; Thanos and Maniatis, 1992). In addition HMG-I(Y) has been implicated as a crucial host protein for HIV-1 viral integration (Farnet and Bushman, 1997), an activity that may be related to observations that fusions of HMG-I(Y) and cellular proteins are associated with specific chromosome translocations frequently found in human lymphomas and leukemias.

Proteins of the HMG-I family are ~10 kDa in size, comprise variable sequences at the N-terminus, an acidic C-terminus, and three short DNA binding domains (DBDs) separated by linkers of 11–23 amino acids. The three DBDs, which have also been termed AT-hooks, recognize a wide variety of A,T rich sequences four to eight base pairs in length (Reeves and Nissen, 1990). We determined the solution structure of a complex between a truncated form of HMG-I(Y), consisting of the second and third DNA binding domains, and a DNA dodecamer containing the PRDII site of the interferon-β promoter (Huth et al., 1997). The stoichiometry of the complex is one molecule of HMG-I(Y) to two molecules of DNA. The structure reveals a new architectural minor groove binding motif which stabilizes B-DNA, thereby facilitating the binding of other transcription factors in the opposing major groove.

In the absence of DNA, the NMR spectrum of HMG-I(2/3) is indicative of a random coil. Upon binding, the two DBDs that contact the DNA become ordered and adopt a well defined conformation in the minor groove. The DNA in the DBD2 and DBD3 complexes is essentially B-type.

The HMG-I DBDs can be divided into three modular components: a central Arg-Gly-Arg core that adopts an extended conformation deep in the minor groove, a pair of lysine and arginine residues at either end of the core that mediate electrostatic and hydrophobic contacts with the DNA backbone, and in the case of DBD2 (but not DBD3) a more extensive network of six amino acids C-terminal to the core that interacts with the sugar-phosphate backbone on either edge of the minor groove. The latter, which results in a significant increase in the DNA contact surface, is the distinguishing feature between the HMG-I type-I (DBD2) and type-II (DBD3) motifs and accounts for the higher DNA binding affinity of the type-I motif. This is illustrated in Fig. 6.

The Arg-Gly-Arg core presents a narrow concave surface which is perfectly suited to insert into the minor groove of A,T-tracts without causing a large perturbation in the DNA conformation . The sidechains of the arginine core residues are oriented parallel to the minor groove and extend away from the central A,T base pair. The conformation of the core arginines is facilitated by the presence of an intervening glycine residue whose CαH protons are packed against the base of an adenine and whose backbone amide is hydrogen bonded to the O2 atom of a thymine. This close proximity of the backbone to the bases precludes any other amino acid at this position. Moreover, the nature of these core interactions with the DNA bases excludes a G·C from the central four base pairs. The large 6-NH$_2$ group of guanine in place of a proton at the equivalent position for adenine would introduce substantial bulk into the narrow minor groove such that the snug fit of the core arginine and glycine methylene groups would be prevented. Trans prolines on either side of the Arg-Gly-Arg core direct the peptide backbone away from the minor groove, and position amino acids N- and C-terminal to the core near the phosphate backbone where a pair of lysine and arginine residues interact with the DNA in a similar manner in the two complexes. Thereafter the interactions of DBD2 and DBD3 with the DNA are different. For the weakly binding DBD3 only one additional lysine residue contacts the DNA, whereas for the strong binding DBD2, a network of additional polar and hydrophobic contacts is observed. A key component of the latter module is a type II turn which positions a lysine sidechain towards a phosphate on the top strand of the DNA and the backbone amide of a glycine close to the adjacent one.

Neither DBD could interact optimally with DNA in either a compressed or expanded minor groove. Hence, it seems likely that the principal architectural role of HMG-I probably involves reversing and preventing intrinsic distortions in DNA conformation, including bending or kinking, by binding in the minor groove, thereby facilitating specific recognition of the opposing major groove by other transcriptional factors.

Interestingly, the structure of the HMG-I DBD2 complexed to DNA comprises features known from polypyrrolecarboxyamide DNA-binding drugs. The Arg-Gly-Arg core mimics the conformation of drugs such as netropsin bound to the minor groove of DNA which is decorated with additional polar and hydrophobic amino acids in the HMG-I DBDs resulting in increased binding affinity.

The modular mode of DNA recognition by HMG-I is reminiscent of that found in a large number of unrelated DNA binding domains. Combination of several DNA binding domains, which may either be very similar, as in the case of classical zinc finger arrays or the DNA binding domains of c-Myb, or different as in the case of the POU domain transcription factors, represent one method of increasing specificity. An alternative approach involves homo- or heterodimerization, as exemplified by the prokaryotic repressors and activators, hormone receptor DNA binding domains, and helix-loop-helix and leucine zipper DNA binding domains. Frequently one or more of the DNA recognition modules are unstructured and highly flexible in the protein alone and only adopt an ordered conformation in the complex. This was observed for the C-and N-terminal tails of the transcription factors GATA-1 and GAGA, and for the entire HMG-I(Y) protein.

CONCLUDING REMARKS

The recent development of a whole range of highly sensitive multidimensional heteronuclear edited and filtered NMR experiments has opened a new area of structure determination by NMR. Protein complexes in the 20–50 kDa range are now amenable to detailed structural analysis in solution.

Despite these advances, it should always be borne in mind that there are a number of key requirements that have to be satisfied to permit a successful structure determination of protein complexes by NMR. The complex in hand must be soluble and should not aggregate up to concentrations of about 0.5–1 mM; it must be stable at room temperature or slightly higher for considerable periods of time (particularly as it may take several months of measurement time to acquire all the necessary NMR data); it should not exhibit significant conformational heterogeneity that could result in extensive line broadening; and finally the protein must be amenable to uniform ^{15}N and ^{13}C labeling. At the present time there are still only relatively few examples in the literature of protein/DNA complexes that have been solved by NMR. One advance that will most likely become a reality in the future is the use of ^{15}N and ^{13}C labeling for the DNA as well. This will allow for easier assignment of DNA resonances, especially in the little dispersed sugar regions of the spectra. One can anticipate, therefore, that over the next couple of years, by the widespread use of multidimensional heteronuclear NMR experiments coupled with semi-automated assignment procedures, many more such structures will become available.

ACKNOWLEDGMENTS

We thank Ad Bax and all past and present members of the combined NMR laboratories at NIH for numerous stimulating discussions. The work in the authors' laboratory was in part supported by the AIDS Targeted Antiviral Program of the Office of the Director of the National Institutes of Health.

REFERENCES

Bax, A. and Grzesiek, S. (1993) *Acc. Chem. Res.* **26**, 131–138.
Bax, A. and Pochapsky. S.S. (1992) *J. Magn. Reson.* **99**, 638–643.
Bax, A., Vuister, G.W., Grzesiek, S., Delaglio, F., Wang, A.C., Tschudin, R. and Zhu, G. (1994) *Methods Enzymol.* **239**, 79–105.
Berg, J.M. & Shi, Y. (1996) *Science* **271**, 1081–1085.
Billeter, M., Qian, Y.Q., Otting, G., Müller, M., Gehring, W. and Wüthrich, K. (1993) *J. Mol. Biol.* **234**, 1084–1097.
Bustin, M. and Reeves, R. (1996) *Progr. Nucl. Acids Res.* **54**, 35–100.
Chuprina, V.P., Rullman, J.A.C., Lamerichs, R.M.N.J., van Boom, J.H., Boelens, R. and Kaptein, R. (1993) *J. Mol. Biol.* **234**, 446–462.
Clore, G.M. and Gronenborn, A.M. (1987) *Protein Eng.* **1**, 275–288.
Clore, G.M. and Gronenborn, A.M. (1991a) *Science* **252**,1390–1399.
Clore, G.M. and Gronenborn, A.M. (1991b) *Progr. Nucl. Magn. Reson. Spectrosc.* **23**, 43–92.
Clore, G.M., Kay, L.E., Bax, A. and Gronenborn, A.M. (1991) *Biochemistry* **30**, 12–18.
Dyson, H.J., Gippert, G.P., Case, D.A., Holmgren, A. and Wright, P.E. (1990) *Biochemistry* **29**, 4129–4136.
Evans, T. and Felsenfeld, G., (1989) *Cell* **58**, 877–885.
Fairall, L., Schwabe, J.W.R., Chapman, L., Finch, J.T. & Rhodes, D. (1993) *Nature* **366**, 483–487.
Farnet, C.M. and Bushman, F.D. (1997) *Cell* **88**, 483 492.
Forman-Kay, J.D., Clore, G.M., Wingfield, P.T. and Gronenborn, A.M. (1991) *Biochemistry* **30**, 2685–2698.
Garrett, D.S., Kuszewski, J., Hancock, T.J., Lodi, P.J., Vuister, G.W., Gronenborn, A.M. and Clore, G.M. (1994) *J. Magn. Reson. Series B* **104**, 99–103.
Goodfellow, P. N., and Lovell-Badge, R. (1993). *Annu. Rev. Genet.* **27**, 71–92.
Granok, H., Leibovitch, B.A., Shaffer, C.D. & Elgin, S.C.R. (1995) *Curr. Biol.* **5**, 238–241.
Griesinger, C., Sørensen, O.W. and Ernst, R.R. (1986) *J. Chem. Phys.* **85**, 6837–6852.
Grzesiek, S., Wingfield, P.T., Stahl, S.J. and Bax, A. (1995) *J. Am. Chem. Soc.* **117**, 9594–9595.
Grzesiek, S. and Bax, A. (1997) *J. Biomol. NMR* **9**, 207–211.

Gustafson, M. L., and Donahoe, P. K. (1994) *Annu. Rev. Med.* **45**, 505–524.

Hannon,R., Evans,T., Felsenfeld, G., Gould, H., (1991) *Proc. Natl. Acad. Sci. U.S.A.* **88**, 3004.

Haqq, C. M., King, C.-Y., Ukiyama, E., Falsafi, S., Haqq, T. N., Donahoe, P. K., and Weiss, M. A. (1994) *Science* **266**, 1494–1500.

Hu, J.-S. and Bax, A. (1997) *J. Biomol. NMR* **9**, 323–328.

Huth, J.R., Bewley, C.A., Nissen, M.S., Evans, J.N.S., Reeves, R., Gronenborn, A.M. & Clore, G.M. (1997) *Nature Struct. Biol.* **4**, 657–665.

Ikura, I. and Bax, A. (1992). *J. Am. Chem. Soc.* **114**, 2433–2440.

Kay, L.E., Keifer, P. and Saarinen, T. (1992) *J. Am. Chem. Soc.* **114**, 10663–10665.

Kay, L.E., Clore, G.M., Bax, A. and Gronenborn, A.M. (1991) *Science* **249**, 411–414.

Kim, Y., Geiger, J.H., Hahn, S. & Sigler, P.B. (1993) *Nature* **365**, 512–520.

Kim, J.L., Nikolov, D.B. & Burley, S.K. (1993) *Nature* **365**, 520–527.

Klug, A. & Schwabe, J.W.R. (1995) *FASEB J.* **9**, 597–604.

Kuszewski, J., Qin, J., Gronenborn, A.M. and Clore, G.M. (1995) *J. Magn. Reson. Series B* **106**, 92–96.

Kuszewski, J., Gronenborn, A.M. and Clore, G.M. (1995) *J. Magn. Reson. Series B* **107**, 293–297.

Kuszewski, J., Gronenborn, A.M. and Clore, G.M. (1996) *Protein Science* **5**, 1067–1080.

Kuszewski, J., Gronenborn, A.M. and Clore, G.M. (1997) *J. Magn. Reson.* **125**, 171–177.

Laudet, V., Stehelin, D., and Clevers, H. (1993). *Nucl. Acids Res.* **21**, 2493–2501.

Love, J.J., Li, X., Case, D.A., Giese, K., Grosschedl, R. & Wright, P.E. (1995) *Nature* **376**, 791–795.

Luisi, B. F., Xu, W., Otwinowski, Z., Freedman, L.P., Yamamoto, K.R. and Sigler, P.B. (1991) *Nature* **352**, 497–505.

Marmorstein, R., Carey, M., Ptashne, M., Harrison, S.C., (1992) *Nature* **356**, 408–414.

Ogata, K., Morikawa, S., Nakamura, H., Sekikawa, A., Inoue, T., Kanai, H., Sarai, A., Ishii, S. and Nishimura, Y. (1994) *Cell* **79**, 639–648.

Omichinski, J.G., Clore, G.M., Schaad, O., Felsenfeld, G., Trainor, C., Appella, E., Stahl, S.J., and Gronenborn, A.M. (1993) *Science* **261**, 438–446.

Omichinski, J.G., Pedone, P.V., Felsenfeld, G., Gronenborn, A.M. and Clore, G.M. (1997) *Nature Struct. Biol.* **4**, 122–132.

Orkin, S. H. (1992) *Blood* **80**, 575–581.

Otting, G. and Wüthrich, K. (1990) *Quart. Rev. Biophys.* **23**, 39–96.

Pavletich, N.P. & Pabo, C.O. (1991) *Science* **252**, 809–816.

Pavletich, N.P. & Pabo, C.O. (1993) *Science* **261**, 1701–1707.

Pedone, P.V., Ghirlando, R., Clore, G.M., Gronenborn, A.M., Felsenfeld, G. & Omichinski, J.G. (1996) *Proc. Natl. Acad. Sci. U.S.A.* **93**, 2822–2826.

Reeves, R. and Nissen, M.S. (1990) *J. Biol. Chem.* **265**, 8573–8582.

Rice, P.A., Yang, S.-W., Mizuuchi, K. & Nash, H.A. (1996) *Cell* **87**, 1295–1306.

Schwabe, J. W. R.,Chapman,L., Finch, J.T. and Rhodes, D. (1993) *Cell* **75**, 567–578.

Sinclair, A.H., Berta, P., Palmer, M.S., Hawkins, J. R., Griffiths, B. L., Smith, M. J., Foster, J. M., Frischauf, A-M., Lovell-Badge, R., and Goodfellow, P. N. (1990) *Nature* **346**, 240–244.

Thanos, D. and Maniatis, T. (1992) *Cell* **71**, 777–789.

Tjandra, N., Garrett, D.S., Gronenborn, A.M., Bax, A. and Clore, G.M. (1997) *Nature Struct. Biol.* **4**, 443–449.

Tjandra, N., Omichinski, J.G., Gronenborn, A.M., Clore, G.M. and Bax, A. (1997) *Nature Struct. Biol.* **4**, 732–738.

Venters, R.A., Metzler, W.J., Spicer, L.D., Mueller, L. and Farmer, B.T. II (1995) *J. Am. Chem. Soc.* **117**, 9592–9593.

Vuister, G.W., Kim, S.-J., Wu, C. and Bax, A. (1994) *J. Am. Chem. Soc.* **116**, 9206–9210.

Werner, M.H., Huth, J.R., Gronenborn, A.M., and Clore, G.M. (1995) *Cell* **81**, 705–714.

Wüthrich, K. (1986) *NMR of Proteins and Nucleic Acids*, Wiley, New York.

Zhang, H., Zhao, D., Revington, M., Lee, W., Jia, X., Arrowsmith, C. and Jardetzky, O. (1994) *J. Mol. Biol.* **238**, 592–614.

2

FITTING PROTEIN STRUCTURES TO EXPERIMENTAL DATA

Lessons from before Your Mother Was Born

Jeffrey C. Hoch, Alan S. Stern, and Peter J. Connolly

Rowland Institute for Science
100 Edwin H. Land Blvd.
Cambridge, Massachusetts 02142

ABSTRACT

For a very long time scientists have been trying to construct models of things they could not directly observe. Plato wrote about this type of problem in his Republic: In the allegory of the cave, he considered the problem of discerning the shape of an object from the shadows it casts in the firelight on the walls of the cave. There is much to be learned from scientists who labored before it was even possible to detect nuclear spin resonance about the proper way to construct physical models that are consistent with indirect observations. In the informal spirit of an Erice afternoon, we will describe the insights of a mathematician, Lagrange, a minister, Bayes, and a biochemist, Haldane, and apply them to protein structure determination by NMR. Some profound insights come from a very unlikely source—a manifesto by Haldane on how to fake data.

INTRODUCTION

The complete determination of a structural model for an n-atom protein in solution requires at least $3n-6$ independent pieces of information. (In x-ray crystallography, where we need to know the position and orientation of the protein within the unit cell, at least $3n$ are needed.) The common amino acids contain on average 16 atoms per residue, requiring over 40 pieces of independent information to determine their relative positions. Modern NMR studies of proteins incorporating ^{13}C and ^{15}N labeling are routinely capable of providing measurements of 20 nuclear Overhauser effects per residue and two vicinal coupling constant measurements per residue. Additional structural information in the form of

Protein Dynamics, Function, and Design, edited by Jardetzky *et al.*
Plenum Press, New York, 1998

hydrogen bonds can be inferred from patterns of NOEs; chemical shifts can also provide structural information. Usually, however, the total number of experimental observations falls well short of the number needed to completely determine a three dimensional structure of the protein. The situation is even more dire than it appears at first glance, for two reasons. One is that the experimental observations in NMR are not independent, but are often quite highly correlated with one another, meaning that some of the information they contain is redundant. Another reason is that proteins are not static, but undergo a wide range of motions. More than $3n$–6 pieces of information are needed to characterize a time-dependent structure.

The solution to both of these dilemmas lies in the fact that we have lots of prior information about proteins that can be used to greatly constrain the range of possible models, even in the absence of sufficient experimental data to completely determine the model. We usually know, for example, the sequence of amino acids that constitute the protein; we know the bonded structures of the amino acids, and how amino acids are bonded to one another. Experience, not to mention theory, shows us that there is relatively little variation in bond lengths or bond angles for a given amino acid, so to a very good approximation we can consider them as part of our prior knowledge. With that assumption, the number of degrees of freedom that need to be defined to construct a model is greatly reduced to just rotation about single bonds, the "soft" degrees of freedom, or dihedral angles. There are three dihedral angles for each amino acid residue that define the conformation of the peptide backbone. But one of these, the omega dihedral angle, is constrained by the partial double-bond character of the C-N peptide bond. On average, the amino acid side chains have two soft dihedral angles. Also left out of this count are dihedral angles that define the orientation of hydrogen atoms but not "heavy" (carbon, nitrogen, oxygen, and sulfur) atoms. Proline represents another special case; although there are five dihedral angles defining its conformation, one also describes part of the peptide chain conformation, and the others are correlated by the constraint that the side chain forms a closed ring. These correlations allow the ring conformation to be described using two parameters: a "pseudoangle" and a "phase" (Haasnoot et al., 1981). The bottom line of this accounting is that we need to define about four degrees of freedom per amino acid residue in order to specify the conformation of the heavy atoms of a protein.

Now the problem of protein structure determination by NMR looks much more manageable: We have plenty of data, unless it is almost completely and diabolically correlated. To determine a structure, we need a practical way to incorporate our prior knowledge about protein structure as we search for models that are consistent with the experimental data. One way to do this would be to use a physical ball-and-stick, or "tinker-toy" model of the protein. Bond lengths would be preserved by the lengths of the sticks, and bond angles would be preserved by drilling holes in the balls separated by the prescribed angles. We would adjust the model by rotating the balls about the sticks in order to satisfy all of the distance and dihedral angle constraints that we are able to derive from the experimental data. This sounds like an absurdly difficult task, and it is—at least when the experimental data describes geometrical relationships between pairs of atoms. In x-ray crystallography, the data contain information on the positions of individual atoms in space, and it was not so long ago that models were fit to electron density maps computed from x-ray diffraction data using just such ball-and-stick models of proteins. A special contraption called a Richards Box was devised to aid the process: it consisted of a partially-silvered mirror at a 45° angle inside a large box, which allowed an image of electron density contour maps to be superimposed directly on the toy protein model (Richards, 1997). Today, of course, models are not manipulated physically, but as computer representations.

A number of different ways have evolved for manipulating computer models of a protein to match experimental NMR data. Two of the most widely used are distance geometry (Kuntz et al., 1989; Havel, 1991) and restrained molecular mechanics (Nilges et al., 1988; van Gunsteren et al., 1989). In distance geometry methods, the conformation of the protein is described by pair-wise distances between atoms. Upper and lower bounds on the distances are inferred from the experimental data and from prior knowledge of the co-valent connectivities of the amino acids. These latter distance restraints are called holonomic bounds. A distance matrix is constructed by randomly choosing, for each pair of atoms, a distance that lies between the upper and lower bounds. The selection of distances is also governed by certain purely geometric considerations such as the triangle inequality: The distance from A to B plus the distance from B to C cannot be shorter than the distance from A directly to C. If all the distance relations can be satisfied simultaneously, then the eigenvectors of the distance matrix corresponding to its three largest eigenvalues describe the cartesian coordinates of the atoms for a model that satisfies the distance relations.

The basis of restrained molecular mechanics methods is an empirical potential energy function describing how well the model matches our prior knowledge of proteins. The function includes bond length, angle, and dihedral terms, but also includes terms to reflect the facts that like charges repel, unlike charges attract, and non-bonded atoms do not approach closer than the sum of their van der Waals radii. Whether by Monte Carlo conformational searching, genetic algorithms, or molecular dynamics simulation, one aims to find the lowest energy conformation—hence the one most compatible with our prior knowledge—that satisfies the experimental constraints. (Note carefully that this "lowest energy" conformation is not necessarily the most stable thermodynamically, because the potential energy function we use is only an approximation to the true internal energy.)

Conceptually, distance geometry and restrained molecular mechanics methods have much in common, but at a detailed level they involve distinctly different computations. It is not unusual to see them used together in a hybrid approach. In the remaining discussion we will adhere to the framework of restrained molecular mechanics, but a similar treatment could be developed for distance geometry methods.

ENTER LAGRANGE

Mathematically we can describe the restrained molecular mechanics approach to structure determination as

$$\text{Minimize } E(\mathbf{x}) \text{ subject to } C(\mathbf{x}) \leq C_0, \tag{1}$$

where $E(\mathbf{x})$ is the empirical potential energy, $C(\mathbf{x})$ is a measure of the agreement between the model described by $\mathbf{x}$ and the experimental data, and C_0 is the experimental uncertainty. This type of problem is called a constrained optimization problem. In general there is no formal analytical solution to this problem, and we are forced to seek numerical solutions. Finding them isn't easy; the condition $C(\mathbf{x}) \leq C_0$ defines a highly complex subregion of conformational space and we want the lowest minimum of $E(\mathbf{x})$ that falls in this subregion. By introducing a variable called a Lagrange multiplier (after Joseph-Louis Lagrange, 1736–1813), the constrained optimization problem can be converted into an equivalent unconstrained optimization problem that is easier to solve. We construct an objective function

$$Q(\mathbf{x}) = E(\mathbf{x}) + \lambda C(\mathbf{x}) \tag{2}$$

where λ is the Lagrange multiplier. This unconstrained problem is solved for the minimum of $Q(\mathbf{x})$; the value of λ is chosen to satisfy $C(\mathbf{x}) \le C_o$. Because $\lambda C(\mathbf{x})$ is added to a function which has units of energy, it is called a pseudo-energy or constraint energy. It is possible to include in $C(\mathbf{x})$ different types of experimental data, for which the errors are not all the same. For the moment we don't need to be very precise about $C(\mathbf{x})$; at the most simple level Eq. (2) will yield some valuable insights.

A minimum of $Q(\mathbf{x})$ will satisfy the condition

$$0 = \frac{\overline{\partial Q}}{\partial \mathbf{x}} = \frac{\overline{\partial E}}{\partial \mathbf{x}} + \lambda \frac{\overline{\partial C}}{\partial \mathbf{x}} . \tag{3}$$

Recalling that the force on an object is given by the negated gradient of the potential energy, we can see that Eq. (3) is satisfied whenever the "force" imposed by the composite energy function on the model is zero. There are two types of solutions to Eq. (3): those with

$$\frac{\overline{\partial E}}{\partial \mathbf{x}} = \frac{\overline{\partial C}}{\partial \mathbf{x}} = 0 \tag{4}$$

and those with

$$\frac{\overline{\partial E}}{\partial \mathbf{x}} = -\lambda \frac{\overline{\partial C}}{\partial \mathbf{x}} \ne 0 . \tag{5}$$

We call the solutions given by Eq. (4) trivial solutions, because the structures are local minima of $E(\mathbf{x})$ alone. Conversely, solutions given by Eq. (5) are nontrivial, and result from the forces on the model due to the empirical potential energy canceling out the forces due to the constraint energy (scaled by λ). In solid state physics, a system for which it is not possible to simultaneously minimize all of the contributions to the total energy is said to be "frustrated"; the analogy is appealing but shouldn't be taken too literally, since the constraint "energy" isn't a real energy, of course.

To put it another way, experimental restraints that have a zero gradient don't exert a force on the model, and restraints that don't exert a force don't constrain the model. This is significant because of a form that is often used for the distance restraint potential corresponding to nuclear Overhauser effect measurements: a flat-well potential. In a flat- or square-well restraining potential, the gradient of the constraint energy is zero if the distance lies between the experimentally determined upper and lower distance bounds. Outside this range the constraint energy increases quadratically with the difference from the nearest bound:

$$C(\mathbf{x}) = \begin{cases} k(r(\mathbf{x}) - r_{max})^2, & r(\mathbf{x}) > r_{max} \\ 0, & r_{min} \le r(\mathbf{x}) \le r_{max} \\ k(r_{min} - r(\mathbf{x}))^2, & r(\mathbf{x}) < r_{min} \end{cases} \tag{6}$$

With a potential of this form, restraints that are satisfied by the model are trivial, and the actual distance is determined completely by the local properties of the empirical potential energy function. Restraints that do constrain the model will be violated, at least slightly, otherwise they would not exert any force on the model.

Would it be safe to conclude that if one performed a series of structure calculations, and in every structure a particular experimental distance restraint is never violated, that the restraint contains no information of relevance to the structure? Almost certainly not. The potential energy surfaces for proteins contain many local minima (the "rugged landscapes" described by Hans Frauenfelder). If an experimental constraint is not violated in any of our final, folded structures, it means only that near the correct local minimum, the potential energy surface (and perhaps other experimental restraints) are sufficient to determine the structure. It could well be that the apparently useless constraint is necessary to *find* the correct local minimum, however. In other words, the constraint might help to determine the overall fold at low resolution, but not the fine details of structure.

To determine if an experimental constraint is truly trivial, we need to look a bit more closely. One way is to monitor constraint violations during the entire structure calculation, not just at the end. If a constraint is never violated at any time during the calculation, then it can be said to be trivial. This may not be a terribly useful criterion, though, if the starting structure involves random atom positions, for which any restraint is likely to be violated. It also won't readily identify correlated or redundant restraints.

Another way to identify trivial restraints is through the use of a reduced representation of protein structure. By mapping the experimental restraints onto the reduced representation, it is quite easy to identify restraints that have no bearing on the overall fold, without performing a structure calculation. A simple representation that we use in our laboratory helps to illustrate the procedure (Hoch and Stern, 1992; Connolly et al., 1994). Consider a protein representation consisting of two pseudo-particles per residue, one representing the α carbon and the other the β carbon (or the first α proton for glycine; see Fig. 1). Experimental distance constraints between atoms can be converted to this representation by a simple procedure, illustrated in Table 1. Each atom is mapped to a pseudo-particle, and the distance restraint between two atoms is converted to a restraint between the corresponding pseudo-particles. For example, both the H^N and H^α protons map to the C^α pseudo-particle, and so restraints involving H^N or H^α in a given residue are converted into restraints involving the C^α pseudo-particle. This conversion requires increasing the upper bound of the distance restraint to reflect the fact that the atoms are usually closer together than the corresponding pseudo-particles are; the further an atom is from its pseudo-particle, the larger the correction factor will be.

Quite often restraints involving totally different pairs of atoms are converted into restraints involving the same pair of pseudo-particles. Only the shortest distance restraint between a particular pair of pseudo-particles needs to be retained, since a longer restraint will never be violated if the shortest is satisfied. At the level of the reduced representation, the longer restraints are trivial.

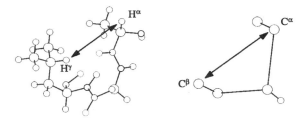

Figure 1. A distance restraint in the all-atom representation is shown at left. On the right is the distance restraint after mapping to the two-particle representation.

Table 1. Mapping of all-atom restraints into two-particle restraints

All-atom restraint	2p restraint	Applied restraint
41 HD# — 9 HB1 4.7	41 CB — 9 CB 7.0	
41 HD# — 9 HB2 4.7	41 CB — 9 CB 7.0	
41 HE# — 9 HB1 5.3	41 CB — 9 CB 8.3	
41 HE# — 9 HB2 5.3	41 CB — 9 CB 8.3	41 CB — 9 CB 7.0
41 HE# — 9 HG# 6.3	41 CB — 9 CB 9.9	
41 HE# — 9 HD# 5.7	41 CB — 9 CB 9.9	
41 HD# — 9 HD# 5.7	41 CB — 9 CB 9.9	

It may seem a bit frustrating that after all the hard work of assigning and quantifying NOEs, only a subset actually contributes to the determination of the structure. Each datum carries information, after all, even though it may not be independent from other experimental observations. There ought to be some way to ensure that all of the data contributes to the determination of the structure. The real culprit in this case is not so much the fact that observations are often correlated, but rather the way in which the experimental constraints are applied; in other words, the flat-well did it. Before we describe a better alternative to the flat-well, we will introduce some pertinent ideas due to Haldane—in a very unlikely guise.

HALDANE'S PRESCRIPTION FOR FAKING DATA

J.B.S. Haldane (1892–1964) was a British scientist (biochemist, physiologist, geneticist, communist, and statistician) who is perhaps best known among those of us concerned about proteins for his mathematical treatment of enzyme kinetics. Haldane gave the following advice (reprinted in its original form by Körner in his treatise on Fourier Analysis, 1988) to anyone who wants to fake experimental results without actually performing the experiments.

If you are going to the trouble to fake data, then you must have some expected or desired value of the result. You want your data set to average to this value, but it is very unlikely that independent measurements of a quantity would yield exactly this result. So you should make sure that your data averages to a value that is not too close to the expected value. To make your data appear even more realistic, Haldane suggests that you make sure that the average of the squared differences from the mean—the χ^2 value—is close to its expected value. Taking these precautions is called first-and second-order faking.

Higher-order faking involves making sure the contributions to χ^2 are distributed properly, including their higher moments (mean cube, mean fourth power, and so on). Haldane suggested that this last step was probably not necessary to avoid detection, but that was before modern computers, which can easily perform the necessary computations.

What does all of this have to do with fitting models to experimental data? You'll agree that experimental data will contain an element of random noise. Results predicted by a model, on the other hand, are not randomly distributed at all; for any given choice of model, the predicted values of observable parameters are fixed (unless, of course, you need to invoke quantum theory to predict your observables). A model which fits all of the available experimental data to a degree much better than the experimental error is said to be "overfitted". To see why such a model is not so hot, consider what would happen if you were to perform the experiment over again. Since experimental errors are random, the new data would give slightly different results. But then the model fit to the original data would not fit the new data nearly so well. If the model is correct, then the deviations between the predictions of the model and the measured values should be distributed in accord with experimental error, for almost any realization of the data. To have a better fit would imply that the model contains information that cannot be justified by the experimental results.

Just as faked data shouldn't be too close to the expected result in order to avoid appearing implausible, a model shouldn't agree too closely with the experimental observations if it is to be plausible. By analogy with Haldane, we can say that ensuring that the value predicted by a model agrees closely (but not too closely) with the experimental average is first-order fitting, ensuring that the average squared deviation between the model's predictions and the experimental values is equal to the variance of the experimental error is second-order fitting, and ensuring that the deviations are properly distributed is higher-order fitting.

APPLYING HALDANE'S PRINCIPLES: BAYESIAN RESTRAINT POTENTIALS

Making sure that the residuals (the differences between the predictions from the model and the actual experimental results) have the correct mean square and are properly distributed is easy to say, but not so obvious how to do. A theorem due to Thomas Bayes (1702–1761) can point us in the right direction. Bayes' theorem can be written as

$$P(\mathbf{m}|\mathbf{d}) \propto P(\mathbf{d}|\mathbf{m}) P(\mathbf{m}) . \tag{7}$$

Expressed in words, this says that the probability $P(\mathbf{m}|\mathbf{d})$ that the model $\mathbf{m}$ is correct after the experimental data $\mathbf{d}$ have been measured is proportional to the probability $P(\mathbf{d}|\mathbf{m})$ that the values $\mathbf{d}$ *would* be observed if $\mathbf{m}$ were correct, multiplied by the probability $P(\mathbf{m})$ that $\mathbf{m}$ is correct before the experimental results are known.

We have a function that tells us how plausible a model is in the absence of any experimental data, the empirical energy function $E(\mathbf{x})$, and we can convert it into a probability via the Boltzmann distribution:

$$P(\mathbf{m}) = \frac{1}{Z} e^{-E(\mathbf{x})/kT} \tag{8}$$

where T is a "temperature factor" that measures the extent we believe that the molecule will fail to match our idealized geometry, and the partition function

$$Z = \int e^{-E(\mathbf{x})/kT} d\mathbf{x} \tag{9}$$

assures that the distribution is properly normalized. The partition function is difficult to compute, since it requires evaluating $E(\mathbf{x})$ over all of conformational space, but fortunately, the constant factor of $1/Z$ will not matter for our purposes.

The other difficulty in computing $P(\mathbf{m})$ is the question of the accuracy of the model. We know, for example, that the empirical energy function is approximate, and that we usually neglect the possibility of motional averaging. Although we know the source of some possible errors in the model, it is usually difficult to know their magnitude and distribution. In Eq. (8), the temperature factor serves to express our uncertainty regarding the errors in the potential function.

The probability of the data given the model, P(**d**|**m**), is related to the experimental error, which we can determine empirically by repeating experiments or by measuring noise levels. If the errors are distributed normally with standard deviation σ, the probability of obtaining a particular datum d is given by the Gaussian distribution

$$P(d) = \frac{1}{\sqrt{2\pi\sigma^2}} e^{-(d-\hat{d})^2/2\sigma^2}$$

(10)

where $\hat{d}$ is the correct value. Note that the datum d can correspond to any experimental observable, such as a nuclear Overhauser enhancement or a coupling constant. Given the model **m** we can calculate what $\hat{d}$ must be, and with an estimate of σ (and assuming the errors are independent), we can compute P(**d**|**m**) as

$$P(\mathbf{d}|\mathbf{m}) \propto \prod_{i=1}^{N} e^{-(d_i - d_i^{calc})^2/2\sigma_i^2} .$$

(11)

Putting together Equations (7), (8), and (11), we see that the probability P(**m**|**d**) of the model **m** given the data **d** is just

$$P(\mathbf{m}|\mathbf{d}) \propto \left[\prod_i e^{-(d_i - d_i^{calc})^2/2\sigma_i^2} \right] e^{-E(\mathbf{x})/kT} .$$

(12)

The negative logarithm of the probability (usually called the *log-likelihood*) is given by

$$-\log P(\mathbf{m}|\mathbf{d}) = E(\mathbf{x})/kT + \sum_i (d_i - d_i^{calc})^2/2\sigma_i^2 + \text{constant} .$$

(13)

The most likely model is the one for which P(**m**|**d**) is largest, i.e., for which Eq. (13) is minimized. This brings us full circle, because we now see that after scaling by kT, Eq. (13) has the same form as Eq. (2). Thus, according to Bayes' theorem, the constraint function $C(\mathbf{x})$ must be equal to the negative logarithm of the distribution function for the experimental errors. In particular, for independent, normally distributed errors with standard deviations σ_i, the constraint function must be given by

$$C(\mathbf{x}) = \sum_i (d_i - d_i^{calc})^2/2\sigma_i^2 .$$

(14)

(The factor kT plays the role of λ in Eq. (2).) For example, if measurements of nuclear Overhauser effects are distributed randomly about a mean value according to a Gaussian distribution, the constraint function that will assure that the residuals are similarly distributed is

$$C(\mathbf{x}) = \sum_{nobs} \left(NOE_{obs} - NOE_{calc} \right)^2 / 2\sigma^2 .$$

(15)

To illustrate the influence of the restraint potential on the distribution of residuals, we computed families of structures for the 58-residue protein BPTI, using restraints derived from the crystal structure (Brookhaven Protein Data Bank entry 4PTI) in a two-stage restrained molecular dynamics protocol (Connolly et al., 1996) using the program X-PLOR (Brünger, 1992). Randomly selecting 50% of the proton pairs that (in the crystal structure) are closer than Å 5 resulted in 697 restraints. One family of structures was computed using a flat-well distance restraint potential; upper and lower distance bounds were obtained by adding 0.5 Å to and subtracting 0.2 Å from the actual distance computed from the crystal structure. The other family was computed using a quadratic NOE restraint potential of the form shown in Eq. 15, in which the NOE_{obs} were calculated from the crystal structure using the isolated spin-pair approximation (and NOE_{calc} from the trial structures in the same way). The force constant was empirically chosen to yield NOE restraint violations comparable to the differences between NOE_{obs} and NOE_{calc} for the structures computed using the flat-well restraint.

Histograms of the differences between the actual distances in the crystal structure and the distances in the computed families of structures are shown in Fig. 2. The distribution of distance errors obtained using the flat-well restraint potential (Fig. 2A) is rather strange and unlikely to represent random experimental errors. In that sense it fails to meet Haldane's criterion for higher-order fitting. The peaks in the distribution near 0.5 and −0.2 are a direct consequence of the edges of the flat region in the restraint potential at the upper and lower distance bounds. By contrast, the distribution of distance errors obtained using the NOE restraint potential (Fig. 2B) is more plausible. Its slightly asymmetric shape reflects the $1/r^6$ dependence of the NOE intensity on the distance.

Figure 3 shows histograms of the differences between the target NOEs (computed from the crystal structure) and the NOEs computed for the two families of structures. To

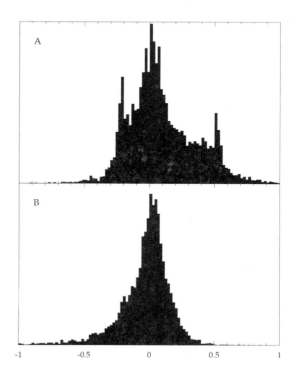

Figure 2. Distribution of distance violations (model–target, in Å) computed from a family of structures obtained using distance (panel **A**) and NOE restraints (panel **B**). Restraints were derived from the crystal structure of BPTI (PDB entry 4PTI). Distance restraints were enforced with a flat-well potential and NOE restraints with a quadratic potential.

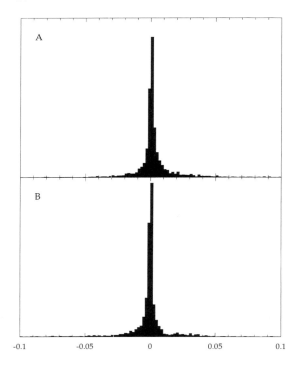

Figure 3. Distribution of NOE violations (computed in the isolated spin-pair approximation, using an isotropic correlation time of 2 nsec and a mixing time of 100 msec) for the families of structures computed using distance (panel **A**) and NOE restraints (panel **B**).

give an idea of the scale, an NOE between a pair of protons 3 Å apart would have an intensity of 0.06 (assuming a correlation time of 2 nsec and a mixing time of 100 msec). The similarity of the distributions indicates that the families of structures do an equally good job of matching the experimentally observable data.

A third measure of the agreement between the computed structures and the target structure is the RMSD of the atomic positions. The flat-well restraint family has an average RMSD of 0.95 Å, and the NOE restraint family has an RMSD of 0.61 Å. This is perhaps understandable, because the flat-well potential carries less information than the NOE potential. In fact, the flat-well potential says nothing at all about interproton distances that fall within the upper and lower bounds. That the distribution in Fig. 2A is not flat between −0.2 and +0.5 Å is because there are other terms in the potential energy function besides the distance restraints. The structures tend to cluster near the local minima of these other energy terms.

In this example we used a very simple theory for the NOE: the isolated two-spin approximation. There are more accurate theories that incorporate multiple-spin effects (Borgias and James, 1989; Gippert et al., 1990), and a number of software packages permit structure refinement using these theories to constrain the model to fit the experimental data. While they are more accurate, these theories are more costly to compute than the isolated two-spin approximation. In the early stages of a refinement the additional accuracy is superfluous, so a reasonable approach is to use an approximate theory to compute a rough model and then use the more costly theory to "polish" the model.

The basic lesson learned from Haldane's prescription applies to other types of experimental data as well. Another widely used source of structural data from NMR experiments is the measurement of vicinal coupling constants, which are related to dihedral angles via the Karplus equation (Karplus, 1963). Often these measurements are used to classify dihedral angles into ranges of values, which are enforced during structure refinement using dihedral angle restraints. Direct refinement against calculated coupling constants, using the Karplus equation and coupling constant restraints, is a better way to satisfy Haldane's prescription, however, especially when conformational averaging occurs. A number of investigators are using this approach (Kim and Prestegard, 1990; Mierke and Kessler, 1993; Garrett et al., 1994). As our understanding of NMR parameters improves, leading to more accurate theoretical description of other types of NMR data, it will be possible to find additional applications of Haldane's prescription. Examples using chemical shift (Case, 1994; Osapay et al., 1994; Oldfield, 1995; Kuszewski et al., 1996), residual dipolar coupling and relaxation effects arising from rotational diffusion anisotropy (Tolman et al., 1997; Tjandra et al., 1997), and cross-correlation effects in magnetic relaxation (Reif and Griesinger, 1997) are already beginning to appear.

CONCLUDING REMARKS

We have seen how principles that were developed long before the advent of NMR spectroscopy apply to the problem of fitting structural models to experimental NMR data. The method introduced by Lagrange for solving problems in constrained optimization shows us how the balance between prior knowledge and experimental data influences the results. Haldane's principles for faking data, when viewed from a slightly different perspective, turn out to be useful for determining when a model is under- or over-fit to experimental data. Bayes' theorem provides the formal link between Haldane's principles and the pseduo-energy or restraint terms used with Lagrange's method to solve the constrained optimization problem. We may summarize by distilling the insights of Lagrange, Haldane, and Bayes to these principles: (1) The deviations between the model and experiment should be distributed in accord with experimental error. (2) The distributions reflect the way in which restraints are applied to the model. (3) Restraints on the values of observables (such as NOE intensities), rather than on derived quantities (such as distances), are the best way to satisfy the first principle.

ACKNOWLEDGMENTS

This work was supported by the Rowland Institute for Science and by grants from the National Institutes of Health (GM-47467) and National Science Foundation (MCB-9316938). We are grateful to Kristin Bartik for many fruitful discussions. We thank David Donoho for introducing us to Haldane's essay on scientific fraud.

REFERENCES

Borgias, B.A., and James, T.L. (1989) *Meth. Enzym.* **176**, 169–183.
Brünger, A.T. (1992) X-PLOR Version 3.1, Yale University Press, New Haven.
Case, D.A. (1994) *Meth. Enzym.* **239**, 392–416.
Connolly, P.J., Stern, A.S., and Hoch, J.C. (1994) *J. Am Chem.Soc.* **116**, 2675–2676.
Connolly, P.J., Stern, A.S., and Hoch, J.C. (1996) *Biochemistry* **35**, 418–426.
Garrett, D.S., Kuszewski, J., Hancock, T.J., Lodi, P.J., Vuister, G.W., Gronenborn, A.M., and Clore, G.M. (1994) *J. Magn. Reson. B* **104**, 99–103.
Gippert, G.P., Yip, P.F., Wright, P.E., and Case, D.A. (1990) *Biochem Pharmacol* **40**, 15–22.

Haasnoot, C.A.G., DeLeeuw, F.A.A.M, deLeeuw, H.P.M., and Altona, C. (1981) *Biopolymers* **20**, 1211–1245.

Havel, T.F. (1991) *Prog. Biophys. Mol. Biol.* **56**, 43–78.

Hoch, J.C., and Stern, A.S. (1992) *J. Biomol. NMR* **2**, 535–543.

Karplus, M. (1963) *J. Am. Chem. Soc.*, **85**, 2870–2871.

Kim, Y., and Prestegard, J.H. (1982) *Proteins* **8**, 377–385.

Körner, T. W. (1988) *Fourier Analysis*, Cambridge University Press, Cambridge, UK.

Kumtz, I.D., Thomason, J.F., and Oshiro, C.M. (1989) *Meth. Enzym.* **177**, 159–204.

Kuszewski, J., Gronenborn, A.M., Clore, G.M. (1996) *J. Magn. Reson. B*, **112**, 79–81.

Merke, D.F., and Kestler, H. (1993) *Biopolymers* **33**, 1003–1017.

Nilges, M., Gronenborn, A.M., Brünger, A.T., and Clore, G.M. (1988) *Protein Eng.* **2**, 27–38.

Oldfield, E. (1995) *J. Biomol. NMR* **5**, 217–225.

Osapay, K., Theriault, Y., Wright, P.E., Case, D.A. (1994) *J. Mol. Biol.* **244**, 183–197.

Reif, B., Hennig, M., and Griesinger, C. (1997) *Science* **276**, 1230–1233.

Richards, F.M. (1997) *Annu. Rev. Biophys. Biomol. Struct.* **26**, 1–25.

Scheek, R.M., van Gunsteren, W.F., and Kaptein, R. (1989) *Meth. Enzym.* **177**, 204–218.

Tjandra, N., Garrett, D.S., Gronenborn, A.M., Bax, A., and Clore, G.M. (1997) *Nat. Struct. Biol.* **4**, 443–449.

Tolman, J.R., Flanagan, J.M., Kennedy, M.A., and Prestegard, J.H. (1997) *Nat. Struct. Biol.* **4**, 292–297.

3

MULTISUBUNIT ALLOSTERIC PROTEINS

William N. Lipscomb

Department of Chemistry and Chemical Biology
Harvard University
12 Oxford Street
Cambridge, Massachusetts 02138

ABSTRACT

Allosteric properties of yeast chorismate mutase, bacterial lactate dehydrogenase, fructose-1,6-bisphosphatase, bacterial phosphofructokinase, glycogen phosphorylase, aspartate transcarbamylase, from *E. coli* and hemoglobin are compared, and some general principles are noted. All show modulation of substrate affinity in the largely rotational movement of subunits. However, the allosteric tetrameric D-3-phosphoglycerate dehydrogenase controls Vmax rather than K_M.

INTRODUCTION

The classic examples of allosteric proteins are multisubunit structures which exhibit T (less active) and R (more active) forms. Affinities of sites for substrates and effectors are influenced by the interactions among these sites, and the resultant control is crucial in control of pathways of metabolism and signaling. The examples discussed here are hemoglobin, aspartate transcarbamylase, glycogen phosphorylase, bacterial phosphofructokinase, fructose-1,6-bisphosphatase, bacterial lactate dehydrogenase, yeast chorismate mutase, and comments on a few others. Earlier summaries have been presented elsewhere (Perutz, 1989; Johnson and Barford,1990; Lipscomb, 1991; Evans, 1991; Stevens and Lipscomb 1993; Mattevi et al., 1996). Although allosteric behavior in single subunit systems is omitted here, attention is called to the structural studies of the allosteric peptide in the activation of trypsinogen (Huber and Bode, 1978). This manuscript contains an attempt to describe guiding principles of allosteric properties (with exceptions) in multisubunit allosteric proteins.

Protein Dynamics, Function, and Design, edited by Jardetzky *et al.*
Plenum Press, New York, 1998

SUMMARY OF CRYSTALLOGRAPHIC FINDINGS

Chorismate Mutase

Chorismate mutase from *Saccharomyces cerevisiae* catalyzes the isomerization of pre-phrenate to chorismate in the biosynthesis of phenylalanine and tyrosine. This enzyme is inhibited by tyrosine and stimulated by tryptophan, and therefore maintains a balance between the separate pathways of chorismate to Tyr (and Phe) as compared with Trp (Braus, 1991).

The allosteric effectors Tyr and Trp bind to the same sites in the T and R forms of the dimeric enzyme (Sträter et al., 1996). Each effector site is about 20Å and 30Å to the active sites. Helix H8 (140–171) connects the allosteric site to the active site which is 30Å away. Information transfer may involve a four-helix bundle: H_2, H_8, H_{11} and H_{12}.

Activation by Trp is due to the steric bulk of its side chain which pushes apart the allosteric domain of one monomer and helix H_8 of the other monomer. Inhibition by Tyr is caused by hydrogen bonds between its OH group and Arg 76 of one monomer and Thr 145 of the other monomer. Phe neither activates nor inhibits. This activation by Trp allows movements of residues which bind to the substrate, especially residues in helices H_2 (Arg 16 and Glu 23), H_8 (Arg 157 and Lys 168), H_{11} (Asn 194 and Glu 198), and H_{12} (Glu 246 and Thr 242) thus increasing the affinity of the enzyme for substrate.

This allosteric transition is characterized by a rotation of one monomer by 15° relative to the other.

Lactate Dehydrogenase

Lactate dehydrogenase from *Bifidobacterium longum* catalyzes the reduction of pyruvate to L-lactate by the enzyme cofactor NADH. Unlike the non-allosteric mammalian enzyme, this bacterial form shows sigmoidal kinetics with pyruvate, and is activated by fructose-1,6-bisphosphate (FBP).

The structure to 2.5Å of a 1:1 complex of T and R forms shows (Iwata et al., 1994) how this tetrameric enzyme controls the substrate affinity by helix sliding between subunits, triggered by FBP. In this T to R transition there are two rotations of subunits; the first is by 3.8° about an axis near His 188, and the second is by 5.8° about the P-axis (Fig. 1). A closed conformation between the dimers related by the P-axis is produced, a

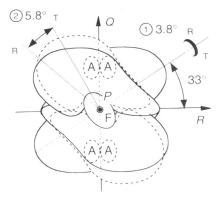

Figure 1. Rotations of subunits in bacterial lactate dehydrogenase (Iwata, et al., 1994).

conformation promoted by FBP which closes the two regions between dimers by binding to and effectively neutralizing positively charged protein residues.

The result of these T to R changes moves the Arg 171 guanidium group 8Å into the active site, and removes His 68 and Ile 240 from the active site. Helix αC (Q) slides about 2.9A at the Cα of His 68(Q), and also causes changes in an "activity controlling loop" and in helix α1/2G which contains Ile 240 (Iwata, et al., 1994). Thus the allosteric transition controls substrate affinity.

Fructose-1,6-Bisphosphatase (FBPase)

Fructose-1,6-bisphosphatase (FBPase) catalyzes the hydrolysis of the 1-phosphate from 2-fructose-1,6-bisphosphate (FBP), aided by a divalent metal ion, Mn^{+2}, Mg^{+2} or Zn^{+2}. This enzyme is inhibited by fructose-2,6-bisphosphate at the active sites and synergistically by AMP at the regulatory sites (Xue et al., 1994). The allosteric transition, caused by AMP, involves a 17° rotation of dimer C_3C_4 relative to C_1C_2. This transition directly affects the binding of one of the two active site divalent cations, but not the affinity of the substrate itself. Strand B3 of the 8-stranded β sheet provides the direct connection between the active site, some 33Å away, to the regulatory site (Zhang et al., 1994).

FBPase from the plant chloroplast is regulated differently: by light-mediated processes, including reduction of disulfide bonds by the ferrodoxin-thioredoxin f system, and by pH changes, divalent cation levels, and fructose-2,6-bisphosphate. The structure of only the reduced form is known (Villeret et al., 1995).

Inhibitors of human FBPase may be candidates for treatment of Type II diabetes.

Phosphofructose Kinase

Phosphofructose kinase from bacteria, (*Escherichia coli and Bacillus stearothermophilis*) uses ATP to make fructose-1,6-bisphosphate from fructose-6-phosphate. The tetrameric bacterial enzyme has an allosteric site per subunit at which phosphoenolpyruvate inhibits and ADP activates. An analogue of this inhibitor is 2-phosphoglycolate which rearranges the 6-F loop at an interface between subunits. This rearrangement causes a 7° rotation of one dimer relative to the other, in the R to T transition, and replaces Arg 162 in the active sites by Glu 161, thereby greatly decreasing the affinity of the enzyme for the negatively charged substrate fructose-6-phosphate (Evans, 1991).

The mammalian enzyme (of the glycolytic pathway) is activated by several molecules, including AMP and fructose-2,6-biphosphate, both of which inhibit the fructose-1,6-bisphosphatase in the gluconeogenic pathway.

Glycogen Phosphorylase (GP)

Glycogen phosphorylase (GP) catalyzes the degradative phophorylation of glycogen with the formation of α-glucose-1-phosphate (G1P), and is regulated by allosteric effectors and by phosphorylation. Muscle GPb, which is unphosphorylated, is inhibited by glucose-6-phosphate and by ATP, and requires AMP for its activity. GPa is phosphorylated at Ser 14, and is active in the absence of AMP.

In the homotropic effects, ligand induced tertiary changes alter the subunit interface of the dimer and, reorient the tower helices (in the region of the dimer interface) which connect to the active site (Johnson and Barford, 1990). The affinity for substrate phosphate in proximity to the phosphate of pyridoxal-5-phosphate is increased in the T to R transition as Asp 283 is displaced by Arg 569. Also, the entry to the active site is facilitated as the loop 281–287 becomes disordered in the T to R transition. During this transition one monomer rotates 10° relative to the other about an axis which is perpendicular to the two-fold axis of the dimer.

In the heterotropic effects, AMP and G6P, and also the phosphorylation of Ser-14 influence the subunit interface of the dimer.

Use of the structure to design strongly bound glucose analogues as inhibitors is known to slow degradation of glycogen, and promote glucose utilization and increased glycogen storage. Such analogues are potential candidates for treatment of Type II diabetes (Johnson, 1997).

Aspartate Transcarbamylase (ATCase)

Aspartate transcarbamylase (ATCase) catalyzes the reaction of carbamyl phosphate with aspartate to yield carbamylaspartate and phosphate, at the very early part of the biosynthetic pathway toward pyrimidines. Allosteric inhibitors are CTP and CTP plus UTP, and the allosteric activator is ATP. The catalytic (c) and regulatory (r) chains are assembled into a dodecamer c_6r_6 of C_3 (almost D_3) symmetry. In the T to R transition the catalytic trimers separate by 11Å and rotate relative to each other by 12°, and the regulatory dimers rotate by 15° around the approximate molecular two fold axis (Lipscomb, 1994). This transition moves Arg 229 into the active site, increasing the affinity for aspartate. Catalysis involves removal of a proton from the NH2 group of aspartate by the phosphate of carbamyl phosphate, as the nitrogen atom of aspartate is carbamylated (Lipscomb, 1994).

The allosteric effectors bind at the regulatory sites, some 60Å away from the nearest active sites. Of the many pathways for communication to the active sites, those involving mutants of the regulatory residues 93–97 (including Lys 94) and Tyr 77 show striking results; the former makes CTP an activator (Wild et al., 1997) and the latter (Tyr 77 to Phe) makes ATP an inhibitor (Van Vliet et al., 1991). Although the homotropic effects and the CTP inhibition approximate the two state model, the ATP activation is independent of the T to R transition (Fetler et al., 1995).

Hemoglobin

Hemoglobin is the first oligomeric allosteric protein for which the T to R transition was structurally characterized (Perutz, 1989). The quaternary change involves a rotation of 15° of an $\alpha_1\beta_1$ dimer relative to the $\alpha_2\beta_2$ dimer when O_2 binds to the heme iron in this tetrameric oligomer. Accompanying this rotation is a shift of 0.8Å in the moveable interface between $\alpha_2\beta_2$ and $\alpha_1\beta_1$ (Perutz, 1989; Baldwin and Chothia, 1979). This shift includes loss of a hydrogen bond between Tyr 42α and Asp 99β in the T form and a gain of a hydrogen bond between Asp 94α and Asn 102β in the R form as the helices repack. A resulting movement of the β subunits toward the central cavity abolishes binding of 2,3-diphosphoglycerate (DPG), the allosteric effector which stabilizes the T form thus promoting release of O_2. Besides DPG, the other specifically bound allosteric effectors are H^+ (Bohr, 1903), and CO_2 which binds to the NH2 terminus of the α chains.

At the tertiary level the T to R transition moves the His-linked Fe about 0.6Å towards the heme plane as O_2 binds. A contact strain between this His and the heme shifts the F6 helix and the FG corner thus propagating the changes through an "allosteric core" (Gelin et al., 1983).

More recent studies have been directed toward the use of mutants to elucidate for the heme-heme interaction multiple pathways which are used either synchronously or synergistically to transmit ligand induced conformational change to a neighboring unligated subunit (Ho). The eight configuration-specific binding intermediates have recently been resolved (Ackers et al., 1992; Huang et al., 1997).

The probable asymmetry of intermediate stages of the R to T transition has been indicated by the measurement of time-resolved resonance Raman spectra of kinetic intermediates following photodissociation of CO from carbonmonoxy hemoglobin (Jayaraman et al., 1995). A crystallographic study of cross-linked hemoglobin also shows the development of asymmetry because the T to R conversion by CO is hampered by the cross-links. In the resulting carbonmonoxy hemoglobin the E helix is in the R state position in the β-subunit, but in the T state position on the α-subunit, whereas the F helix shows the reverse effect (Schumacher, 1995).

Other Allosteric Protein Structures

Other allosteric protein structures (Mattevi et al., 1996) are referred to briefly here. The first is the tetrameric enzyme D-3-phosphoglycerate dehydrogenase which, unlike the examples given above, affects Vmax rather than the effective binding related to K_M, (Grant et al., 1996). Allosteric regulation is caused by binding of L-serine, the product of the metabolic pathway, at a regulatory site which is some 33Å from the ring of the nearest NADH cofactor. The effector Ser is bound near the assembly of two four-stranded β-sheets from two regulatory domains. Also, the active site is shared between two other domains, closure of which is most probably inhibited by the binding of Ser to the regulatory domains. A model of domain movements has been proposed in which these well-structured domains move about flexible hinges in the allosteric transition (Grant et al., 1996). The reported structure is that of the less active T form, from which the R form (with a more closed active site) has been modeled.

In the presence of divalent and monovalent metal ions, pyruvate kinase converts phosphoenolpyruvate (PEP) to pyruvate as ATP is formed from ADP. This is the final reaction in glycolysis, and the tetrameric eukaryotic enzyme normally shows cooperative binding of PEP and allosteric activation by fructose-1,6-bisphosphate. On the other hand, the mammalian muscle enzyme shows no cooperativity, and its structure (Muirhead et al., 1986; Larsen, 1994) most likely corresponds to the R state of the allosteric enzyme (Walker et al., 1992). However, a T state structure of the type 1 *Escherichia coli* isoenzyme which is activated allosterically by fructose-1,6-bisphosphate and inhibited by ATP has been determined to 2.5Å resolution (Mattevi, 1995). The R state of this *E. coli* enzyme is not yet available. Nevertheless, the muscle (M1) enzyme and the *E. coli* T state are compared (Mattevi, 1995). There are reorientations of over 16° of the four subunits within the tetramers, and reorientations of domains B and C by 17° and 15°, respectively, within subunits of the M1 enzyme as compared with the T enzyme. The domain (modular) nature of the movements may become a recurring theme among larger allosteric systems. Connection between the subunit rotations and loop 6 of the $(\beta/\alpha)_8$ barrel distorts the active site of the T form. The structure of the R state of the *E. coli* enzyme and the heterotropic action of fructose-1,6-bisphosphate are subjects for future studies.

NEGATIVE COOPERATIVITY

In positive cooperativity the binding of the first molecule makes easier the binding of the second (Bohr and Hasselbach, 1904; Monod et al., 1995). In negative cooperativity the binding of the second molecule is made more difficult (Koshland et al., 1966; Levitski and Koshland, 1969; Koshland, Jr., 1996), thus enhancing the binding of the first molecule and increasing the concentration range of the response (Koshland, Jr., 1996). Negative co-operativity occurs in several receptors (GABA, insulin, aspartate, estrogen and beta adren-ergic receptors), and in many enzymes (alkaline phosphatase, aspartate aminotransferase, aspartate transcarbamylase, F_1-ATPase, glyceraldehyde-3-phosphate dehydrogenase, glyc-erol dehydratase, glutamate dehydrogenase, malic enzyme, phenylalamine ammonium lyase, pyruvate kinase, 3', 5' cAMP phosphodiesterase, prostaglandin E stimulated adeny-late cyclase, and yeast hexokinase (Koshland, Jr., 1996)).

Negative cooperativity is accommodated by the sequential model (Koshland, Jr. et al. 1966), and most often is seen in a pair of sites as lowered symmetry when only one of the two sites is occupied. The loss of symmetry makes negative cooperativity incompat-ible with the MWC symmetry model (Monod et al., 1965).

In aspartate transcarbamylase the binding of CTP shows three strongly bound sites and three more weakly bound sites. The symmetry, already C_3 instead of D_3 in the unli-gated enzyme, becomes even more asymmetric (within C_3) when CTP binds; thus it be-comes more difficult for the substrate to transform this more asymmetric T state into the R state.

TORQUES IN ALLOSTERIC ENZYMES

Given that most allosteric transitions in oligomers involve relative rotations of subunits, it is tempting to go beyond forces to look at torques, the outer product of the po-sition vector and the force vector. (The force vector is the negative gradient of the poten-tial energy). Of course, for any state at equilibrium torques must be balanced by counter torques, which may also involve the protein or the solvent and its constituents. Neverthe-less, it is possible to assign contributions to torques from individual side chains and to other structural features, including bound molecules.

ATCase

Torques were evaluated from ATP ligated T and R forms, and from CTP ligated T and R forms (PDB 4AT1, 7AT1, 5AT1, 8AT1) after optimization using 20 cycles of Powell minimization. Both Van der Waals and electrostatic (including hydrogen bonding) energies were calculated involving the interactions between effectors and the protein atoms. A posi-tive torque about the two-fold axis (X) and a negative torque about the three fold axis (Z) act in the T to R transition direction. For ATP ligated ATCase, the torque (T to R) about the two fold axis of the T form is mainly due to the Van der Waals interaction of Lys 94r, whereas the torques about the three-fold axis arise mainly from Ala 11r and Ile 12r. Torques which occur when CTP binds are much less. Lys 94r binds to the phosphates and Ile 12r interacts with the base of the effectors. Contributions of ATP and CTP to torques in the looser R state as it converts to the tighter T state are somewhat smaller.

FBPase

In this tetrameric enzyme the binding of the allosteric inhibitor AMP produces a ro-tation around the X axis of rotation in the R to T direction when torques on the C_1C_2 dimer are negative and torques on C_3C_4 are positive. The results indicate that binding of AMP to the C1 monomer gives rise to large negative torques on the following residues: Gly 26,

Thr 27, Glu 29, Lys 112 Tyr 113 and Arg 140, all of which interact with AMP. The largest negative torques arise from interactions with the phosphate of AMP, and the next largest with its ribose oxygens.

Torques were also examined when product fructose-6-phosphate, substrate fructose-1,6-bisphosphate, and active site inhibitor fructose-2,6 bisphosphate were bound. Binding of each of these compounds to the active site of C_1 yielded positive torques which stabilize the R quaternary state. On the other hand when each of these three compounds was bound to the regulatory site the negative torques indicated that the R to T transition is favored.

Further Questions

There are many unanswered questions in these results on torques, including the decoupling of torques and counter torques, the role of solvent and solutes, the eternal question of the effective dielectric constants, the entropy-enthalpy compensation effects, dynamic processes, and no doubt other questions.

PRINCIPLES FOR ALLOSTERIC BEHAVIOR (LIPSCOMB, 1991)

1. Although other structural aspects of control (e.g. dissociation, genetic regulation, bond change) are also prevalent, the enzymes discussed here all rely on oligomerization in which the subunits undergo relative rotational and translational shifts in order to achieve either homotropic (substrate only) or heterotropic regulation.

2. Relative rotations of subunits are present in all of these enzymes. In ATCase (c_6r_6) a c_3 unit rotates 12° relative to the other c_3, and each r_2 rotates 15° in the T to R transition. (The signal distance from the regulatory site to the nearest active site is 60Å.) In FBPase (α_4), one α_2 rotates 17° (signal distance 33Å). In phosphofructokinase (α_4), one α_2 rotates by 7°. In GP (α_2) one monomer rotates 10° (signal distance 30Å). In Hb ($\alpha_2\beta_2$) one $\alpha\beta$ rotates 15° (signal distance 20Å). In AlloLDH(α_4) rotations of 5.8° about a dyad axis and by 3.8° about another axis (signal distance about 25Å). In chorismate mutase one monomer rotates 15° (signal distance 30Å).

3. Relative translations are about 1Å or so in all of these enzymes, except for the increase of 11Å between the c_3's in ATCase, and for the shift of 3Å of the helix containing His 68 in AlloLDH.

4. Information transfer over long distances requires organized secondary and tertiary structure. Many more studies are required to provide structural mechanisms and relate them to the biochemistry and molecular biology of these systems.

5. Secondary structure is preserved, except for one turn of helix in PFK (155–162) and the N-terminal helix which contains the phosphorylated Ser 14 in GPa.

6. The symmetry before and after the allosteric transition is largely preserved. However, partial occupancy of substrate or effectors may lower symmetry.

7. The quaternary transition changes the affinity for substrate. In ATCase Arg 229 is moved. In FBPase metal binding residues are moved, especially Asp 121. In PFK Arg 162 is exchanged for Glu 161. In GP Arg 569 is exchanged for Asp 283. In AlloLDH Arg 171 is exchanged for His 68 of an adjacent chain. In Hb the Fe^{+2} is allowed to move into the plane of the heme by movements of the F

helix and the FG corner. An exception is D-3-phosphoglycerate dehydrogenase in which Vmax is altered.

8. Ligand sites are shared between adjacent subunits in ATCase, FBPase, PFK and AlloLDH. The sharing of sites tends to favor concerted transitions.

9. Apparent partial block of entry of substrates into the active sites occurs in the T forms of Hb, GP, PFK and AlloLDH. However, the flexibility of protein structures may modify the effectiveness of this blockage.

10. Inhibitors and activators share parts of a common site in ATCase (ATP and CTP), in bacterial PFK (phosphoglycollate and ADP), and GPb where G6P and AMP share the phosphate site.

11. Reduction of activity occurs to achieve allosteric control occurs only in Hb (compare myoglobin with hemoglobin) and in ATCase (compare c_3 with c_6r_6).

12. Substrate induces the T to R transition in Hb, ATCase and GP, whereas the allosteric effector induces the R to T transition in FBPase, PFK and AlloLDH.

13. There are also single subunit proteins which show allosteric behavior, especially those that have two (or more) distinct states, and some of which respond to an allosteric ligand, e.g. trypsin (Huber and Bode, 1978).

ACKNOWLEDGMENT

The author thanks the organizers of this International School, and the National Institutes of Health (GM06920) for support. The author wishes to acknowledge the contributions of J.-Y. Liang and Y. Xue to the material on torques.

REFERENCES

Ackers, G. K., Doyle, M. L., Myers, O. and Daugherty, M. W. (1992) *Science* **255**, 54–63.
Baldwin, J. and Chothia, C. (1979) *J. Mol. Biol.* **129**, 175–220.
Bohr, C., Hasselbach, K. A. and Krogh, A. (1904) *Skand. Arch. Physiol.* **16**, 402–412
Bohr, C. (1903) *Zentr. Physiol.* **17**, 682.
Braus, G.H. (1991) *Microbiol. Rev.* **55**, 349–370.
Evans, P. R. (1991) *Current Opinion in Structural Biology* **1**, 773–779.
Fetler, L., Tauc, P., Hervé, G., Moody, M. F. and Vachette, P. (1995) *J. Mol. Biol.* **251**, 243–255.
Gelin, K. H., Lee, A. W.-M., and Karplus, M. (1983) *J. Mol. Biol.* **171**, 489–559.
Grant, A. G., Schuller, D. J. and Banazak, L. J. (1996) *Protein Science* **5**, 34–41.
Ho, C. (1992) *Adv. Protein Chem.* **43**, 153–312.
Huang, J., Koestner, M. L. and Ackers, G. K. (1997) *Biophysical Chemistry* **64**, 157–173.
Huber, R. and Bode, W. (1978) *Accounts Chem. Res.* **11**, 114–122.
Iwata, S., Kamata, K., Yoshida, S., Minowa, T. and Ohta, T. (1994) *Nature Structural Biol.* **1**, 176–185.
Jayaraman, V., Rodgers, K. R., Mukerji, I. and Spiro, T. G. (1995) *Science* **269**, 1843–1848.
Johnson, L. N. and Barford, D. Jr. (1990) *J. Biol. Chem.* **265**, 2409–2412.
Johnson, L. N. (1996) in *Protein Structure-Function Relationships*, Eds: Z. H. Zaidi and D. L. Smith, Plenum, New York, N.Y. pp. 97–108.
Koshland, Jr., D. E. (1996) *Current Opinion in Structural Biology* **6**, 757–761.
Koshland, Jr., D. E. Nemethy, G. and Filmer, D. (1966) *Biochemistry* **5**, 365–385.
Larsen, T. M., Laughlin, L. T., Holden, H. M., Rayment, I. and Reed, G. H. (1994) *Biochemistry* **33**, 6601–6309.
Levitski, A. and Koshland, D. E. (1969) *Proc. Natl. Acad. Sci. USA* **62**, 1121–1128.
Lipscomb, W. N. (1994) *Adv. Enzymol.* **68**, 67–151.
Lipscomb, W. N. (1991) *Chemtracts-Biochemistry and Molecular Biology* **2**, 1–15.
Liu, L., Wales, M. E. and Wild, R. (1997) *Bichemistry* **36**, 3126–3132.
Mattevi, A, Rizzi, M. and Bolognesi, M. (1996) *Current Opinion in Structural Biol.* **6**, 824–829.

Mattevi, A., Valentini, G., Rizzi, M., Speranza, M. L., Bolognesi, M., and Coda, A. (1995) *Structure* **3**, 729–741.

Monod, J., Wyman, J. and Changeux, J. P. (1965) *J. Mol. Biol.* **12**, 88–118.

Muirhead, H. and Schmidt, W., (1986) *EMBO J.* **5**, 475–481.

Perutz, M. F. (1989) *Mechanism of Cooperativity and Allosteric Regulation in Proteins*, Cambridge University Press, Cambridge, England.

Perutz, M. F. (1980) *Proc. Roy. Soc. London* **B208**, 135–162.

Schirmer, T. and Evans, P. R. (1990) *Nature* **343**, 140–145.

Schumacker, M. A., Dixon, M. M., Kluger, R., Jones, R. D. and Brennan, R. G. (1995) *Nature* **375**, 84–87.

Stevens, R. C. and Lipscomb, W. N. (1993) in *Molecular Structures in Biology*, Eds: R. Diamond, T. F. Koetzle, K. Prout and J. Richardson, Oxford University Press, Oxford, UK, pp. 223–259.

Sträter, N., Håkanson, K., Schnappauf, G., Braus, G. and Lipscomb, W. N. (1996) *Proc. Natl. Acad. Sci. USA* **93**, 3330–3334.

Van Vliet, F., Xi, X.-G., De Staercke, C., Wannamaker, B., Jacobs, A., Cherfils, J., Ladjimi, M. M., Hervé, G. and Cunin, R. (1991) *Proc. Natl. Acad. Sci. USA* **88**, 9180–9183.

Villeret, V., Huang, S., Zhang, Y., Xue, Y. and Lipscomb, W. N. (1995) *Biochemistry* **34**, 4299–4306.

Walker, D., Chia, W. N. and Muirhead, H. (1992) *J. Mol. Biol.* **228**, 265–276.

Wild, J., private communication.

Xue, Y. Huang, S., Liang, J.-Y., Zhang, Y. and Lipscomb, W. N. (1994) *Proc. Natl. Acad. Sci. USA* **91**, 12482–12486.

Zhang, Y., Liang, J.-Y., Huang, S. and Lipscomb, W. N. (1994) *J. Mol. Biol.* **244**, 609–624.

4

STUDYING PROTEIN STRUCTURE AND FUNCTION BY DIRECTED EVOLUTION

Examples with Engineered Antibodies

Andreas Plückthun*

Biochemisches Institut
Universität Zürich
Winterthurerstr. 190, CH-8057 Zürich, Switzerland

ABSTRACT

Directed molecular evolution is a powerful strategy for investigating the structure and function of proteins. When a function, such as ligand binding, can only be carried out by the native state of the protein, the biological selection for this function can be used to improve structural properties of the protein. Thus, thermodynamic stability and folding efficiency, which is the ability to avoid aggregation during folding, can be optimized. Three methods of selection are reviewed: phage display, selectively infective phages (SIP) and ribosome display, a cell-free method. Examples for optimizing antibody stability are discussed. In one case, antibodies have been generated under evolutionary pressure, which are stable in the absence of any disulfide bond, in the other case, a kink in the first strand of the beta-sandwich of kappa domains has been optimized.

1. INTRODUCTION

The use of molecular evolution technologies is starting to provide a new perspective in the study of the structure, stability and function of proteins. Representative examples in several application areas such as enzyme activity or antibody affinity illustrate the power of this approach (Hall and Knowles, 1976; Stemmer, 1994a; Low et al., 1996; Baca et al., 1997; Hanes and Plückthun, 1997; Moore et al., 1997). Like the natural evolution of pro-

* Tel: (+41-1) 635 5570; FAX: (+41-1) 635 5712; E-mail: plueckthun@biocfebs.unizh.ch

Protein Dynamics, Function, and Design, edited by Jardetzky *et al.*
Plenum Press, New York, 1998

teins, directed evolution alternates between creation of diversity and selection. Depending on the problem at hand, diversity may be focused on certain regions in a protein, or the whole sequence may be sprinkled with mutations.

Out of this diversity, superior molecules need to be selected. In most non-enzyme cases, the only directly selectable function is binding to a ligand, but many structural features which affect the amount of the protein molecule produced or its stability indirectly affect selection efficiency, as long as ligand binding is limited to the correctly folded molecule. It is this coupling of function to the native state which has been exploited to optimize protein structure in the examples reviewed here.

While the evolutionary approach is only one of several genetic approaches for studying protein structure and function, it has provided a new line of attack for the elucidation of effects which are based on exquisitely complex interactions between amino acids, each contributing only small energies to stability, ligand binding or transition state stabilization. Thus, the technology of directed molecular evolution provides a new sampling mechanism of variants along a pathway of improved function.

In the past, the combination of site directed mutagenesis and biophysical investigations has provided enormous insight into the architecture and design of proteins, by allowing the investigator to change specific amino acids and testing their influence on the parameter of interest, be it stability, folding kinetics, or functional properties of the protein. While very successful for testing hypotheses of the function of a small number of residues, it is only efficient as long as a very good theoretical understanding of the problem under investigation is already available, such that relevant mutants can be planned. Examples where site directed mutagenesis has been instrumental is in the study of enzyme mechanisms, where from a knowledge of the structure those amino acids can often be pinpointed which might be involved in the mechanism (Fersht and Wells, 1991). An important extension is the use of multiple mutants to elucidate synergistic effects—the so-called double mutant cycles (Horovitz, 1996). In this strategy, the relative contribution of one mutant in the presence or absence of another is quantitatively compared, and the relative interaction energy of the two amino acids can be elucidated.

A very systematic mutagenesis of every single residue and multiple combinations thereof can be carried out to define e.g. extended binding sites. This has frequently been called "alanine scanning", since alanine is the ideal residue to which to change a particular residue, alanine being compatible with most secondary structures, sufficiently hydrophilic and not carrying a functional group in the side chain (Ward et al., 1990; Wells et al., 1993). This strategy has shown great success in e.g. defining hormone-receptor interactions or antibody-antigen interactions (Jin et al., 1992; Wells et al., 1993).

The use of such directed mutagenesis approaches has also been successful in the study of protein stability and the temporal events of protein folding (Fersht, 1995; 1997). While very informative, these experiments are quite labor intensive and are not very efficient in uncovering new relations in protein stability and internal interactions, unless carried out on a very large scale.

The main problem in all site-directed mutagenesis strategies is the combinatorial explosion of mutants one might be tempted to make. Consequently, at the present level of understanding, the rational "improvement" of ligand binding or protein stability by cycles of directed mutagenesis, their analysis and deduction of further mutations has not been very successful, and this strategy of the "protein design cycle" has the unfortunate property of being both labor intensive and slow. It is this problem that the technology of "directed evolution" may help to solve, by providing a highly efficient sampling pathway of informative mutants.

2. THE CHALLENGES IN DIRECTED MOLECULAR EVOLUTION

In the strategy of directed molecular evolution nature is imitated by using a succession of mutagenesis and selection. Thus, there are three technical challenges to overcome.

The first is to generate a sufficient number of diverse mutants. While, over the last few years, a number of random mutagenesis techniques have been developed (see below), this problem is far from trivial. Mutations must evenly cover the region of interest, and not be biased for purely technical reasons. Favorable and unfavorable mutations may become covalently linked on the same genome, and obscure the effect of the beneficial mutation and prevent the further selection of this variant. Furthermore, the size of a library can be too small to comprehensively test all variants. It is important to realize that the library size is given not by the chemical efficiency of making a DNA library, but by providing such a library in a screenable format, which usually involves bringing it into bacterial cells. Unfortunately, most of the diversity is lost in this last step, but a solution to this problem will be discussed below.

The second challenge is to devise efficient selection techniques. A number of currently available screening technologies have been reviewed (Phizicky and Fields, 1995) and some are discussed in more detail below. In the context of using molecular evolution for understanding protein structure, folding and mobility, an indirect strategy has to be chosen: the quantity of interest has to be coupled to a selectable property of the protein. For example, ligand binding can be used as a selectable property as long as it depends on the protein being in the native state. Thus, all molecular quantities which influence the number of molecules which reach the native state (e.g. the thermodynamic stability of the protein) can be selected, by enriching those variants which outcompete the others in ligand binding. In other words, those mutants will win the race of which a higher percentage of molecules reaches or remains in the native state—as long as a direct change of the binding pocket can be excluded.

The third challenge is to make the selection strategies so simple that they can be carried out over several generations (usually called "rounds"), that is cycles of mutagenesis and selection. Multiple cycles are necessary, since it is statistically unlikely to find a collection of favorable mutations in a single round, and back-of-the-envelope calculations immediately show that it is not possible to ever make libraries large enough to contain all possible multiple mutations of a protein. Thus, a stepwise approach is needed, just as nature uses, and hence, the cycles of mutagenesis and selection must be very convenient to carry out, in order to allow true evolution over many generations.

3. LOCALIZED AND WHOLE-GENE RANDOM MUTAGENESIS

As detailed above, the evolutionary strategy consists of an alternation between random mutagenesis and phenotypic selection for the property of interest. In this section, a few random mutagenesis strategies will be briefly reviewed. We can first distinguish whether the mutagenesis is completely undirected and scattered all over the protein, or focused to particular regions in the gene (Fig. 1). The two experimental examples given in the last section will illustrate both strategies.

Which one to use obviously depends on whether the phenotypic effect is controlled by a local stretch of sequence or the global structure of the protein. Antibody combining sites (Fig. 2) are a good example for the difficulty in answering this question. Numerous studies have shown that, as one would expect from the 3D structure (Padlan, 1996), the

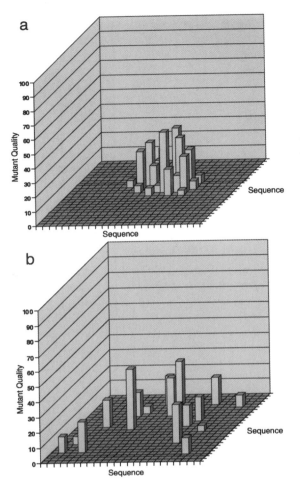

Figure 1. Localized and non-localized mutagenesis. Very schematically, two hypothetical cases are shown, how double mutations in a protein may influence its function (denoted as mutant quality in the z-axis). In (**a**), it is assumed that only a small region in the middle of the sequence is responsible for improving the function. The height of the bars symbolizes the effect of mutations at a particular position in the sequence, and only double mutations are shown (the two sequence axes). All mutations which improve function are localized in the central part of the sequence, and thus the best strategy is clearly to localize mutagenesis to this region. In (**b**), mutations all over the sequence can improve the function, and are correlated in complicated ways, that is mutations in the left part of the sequence can positively interact with those in the right part of the sequence, giving rise to bars scattered all over the 2D map. Importantly, for technical reasons, the total number of bars is about the same (representing the number of mutants which can be experimentally examined, because of restrictions in library size). Thus, in (**b**), an undirected random sampling of mutants is the better strategy than would be the comprehensive sampling of a small sequence region.

residues directly contacting the antigen, which are located in the complementary determining regions (CDRs), play the major role in providing binding energy. However, additional residues, contacting the CDRs, but not the antigen, can further improve binding by subtly changing the conformation of the CDRs. It is this "second sphere" of amino acids, whose contribution to binding is—unfortunately—beyond theoretical prediction today, as the effects are small, indirect and all interdependent on each other. Thus, antibody combining sites may be optimized in two steps: first in a "coarse" adjustment, by selecting the CDR sequences, and second in a "fine tuning", by optimizing their conformation and positioning in space, i.e., by optimizing the "second sphere". It may come as no surprise that this is exactly what happens in the affinity maturation of the antibody during the maturation of the B-cell in the immune response (Wagner and Neuberger, 1996).

In reproducing these mutations *in vitro*, one can also use localized and complete mutagenesis of the gene in question. One of the most effective methods of localized mutagenesis is to replace the CDRs by cassettes made from synthetic oligonucleotides which encode all the variability desired. Usually, phosphoramidites of mononucleotides are the synthetic building blocks for oligonucleotides. In this case, however, the mixture of codons arising from a number of mixed mononucleotides cannot be precisely controlled.

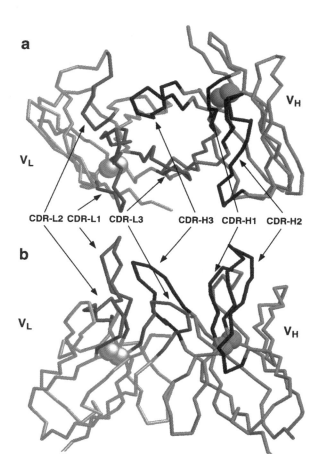

Figure 2. Structure of antibody variable domains, exemplified by the phosphorylcholine binding antibody McPC603 (PDB file 2mcp). The domains and the complementarity determining regions are labeled. The disulfide bonds in the interior of each domain are labeled. (**a**) view from the perspective of the antigen, (**b**) view from the side.

The solution consists in presynthesizing the codons for all 20 amino acids in a format compatible with a modern synthesizer (Virnekäs et al., 1994; Kayushin et al., 1996) (Fig. 3a). This means that the building blocks are phosphoramidites of whole codons and carry protecting groups just as mononucleotides would (Fig. 3a). These building blocks can then be mixed in any ratio, and only those trinucleotides can be added during the synthesis of a particular codon which are desired for the mixture of amino acids to be encoded (Fig. 3b).

To randomize whole genes, in contrast, PCR based methods are most convenient, as they are completely general and can be used in any gene. Many methods have been described for "error prone PCR" (Cadwell and Joyce, 1994), that is conditions in which the error rate is higher than normal. These methods suffer from the inherent disadvantage that either the error rate is too low and the majority of molecules will still be wild-type, or the error rate is too high, and each individual molecule is likely to contain beneficial and deleterious mutations at the same time (Fig. 3c). If several beneficial mutations are required in order to detectably improve the phenotype, it is likely that such molecules will never be found, as they always occur in the context of deleterious mutations, which are probably far more frequent.

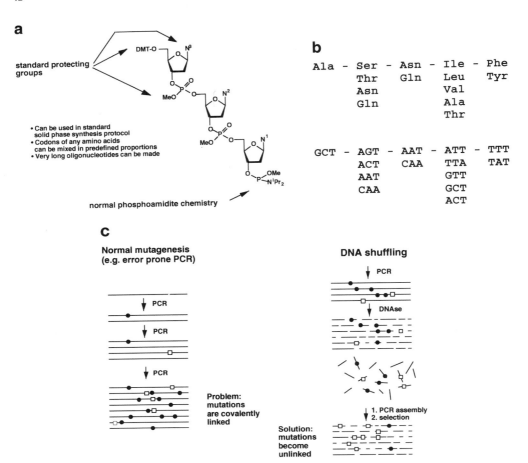

Figure 3. Mutagenesis strategies for localized and whole-gene mutagenesis. (**a**) Trinucleotide building blocks for use in oligonucleotide synthesis. Such presynthesized trinucleotides, representing codons, can be used in standard solid-phase oligonucleotide synthesis. The details can be found in Virnekäs et al., (1994). The chemical structure of the trinucleotide building block is shown, and N^1, N^2 and N^3 are bases. (**b**) Use of the trinucleotides in designing a gene with mixed codons. If a gene with the variability given in the top is desired, the codons given in the bottom need to be mixed during synthesis, using the building blocks given in (a). These mixed codon oligonucleotides are useful in the localized mutagenesis strategy schematically shown in Fig. 1a. (**c**) Mutagenesis strategies useful for undirected mutagenesis, sprinkling mutations over a whole gene of interest. On the left, error prone PCR (see text) is shown schematically. Two types of mutations are shown, favorable ones (open squares) and unfavorable ones (closed circles). In successive cycles of PCR, more mutations of each are introduced, and usually molecules will contain some of either type. Thus, the beneficial effect of the favorable mutations can be completely obscured by the presence of unfavorable ones. On the right, DNA shuffling according to Stemmer (1994a) is shown. A DNase step breaks up the DNA to small pieces, and PCR is used to reassemble the gene. Thereby, mutations are crossed, and genes with largely favorable mutations will be obtained which can be enriched by selection.

Stemmer (1994a) described an elegant solution to this dilemma. By carrying out PCR with an enzyme without proofreading functions, such as the *Taq* polymerase from *Thermus aquaticus*, a certain number of errors are introduced. By then digesting the gene with DNAse into small pieces and PCR assembling the gene from the pieces, the mutations are "crossed", just as if they were located on individual chromosomes (Fig. 3c), even though they belong to the same open reading frame of the protein. Thereby, beneficial mutations can be isolated from deleterious ones by selection (next section), and evolution is much faster than with other methods (Stemmer, 1994a,b).

4. SELECTION TECHNOLOGIES

4.1. General

The common requirement of all selection techniques is to provide such a linkage and physically connect the phenotype with the genotype of the protein (Phizicky and Fields, 1995). That is, if a particular protein molecule has more desirable properties than the rest, because it carries one or more useful mutations, there must be a way of amplifying and immortalizing this particular sequence. Hence, one must achieve a 1:1 physical linkage of the DNA sequence (the genotype, which can be inherited) and the protein sequence (the phenotype, which provides the properties). If there was a pool of protein molecules and a pool of DNA sequences, but no such direct physical linkage, there would be no way that the particular DNA sequence belonging to the improved protein could be identified and amplified.

A whole array of techniques has been developed to select for molecular interactions, such as phage display (Smith, 1985; Dunn, 1996), the yeast two-hybrid system (Bai and Elledge, 1996; Warbrick, 1997), the peptides-on-plasmids system (Cull et al., 1992), ribosome display (Mattheakis et al., 1994; Hanes and Plückthun, 1997), yeast surface display (Boder and Wittrup, 1997) and bacterial display (Georgiou et al., 1997). To select for structural features, however, the selection has to be quite resistant to enriching "sticky" molecules and must be convenient enough to be applied over multiple rounds, in order to enhance very small selective advantages. In this article, three methods will be discussed in more detail which have been shown to be particularly useful in the field of molecular evolution, namely phage display, selectively infective phages and ribosome display.

4.2. Phage Display

The most popular technique used in this field today is phage display (Smith, 1985; Scott and Smith, 1990; Winter et al., 1994; Dunn, 1996; Cortese et al., 1996). It relies on fusing the protein of interest to the minor coat protein of the phage, the gene3 protein (g3p). This protein consists of three domains of 68 (N1), 131 (N2) and 150 (CT) amino acids, connected by glycine-rich linkers of 18 (G1) and 39 (G2) amino acids, respectively (Fig. 4). The first N-terminal domain, N1, is thought to be involved in penetration of the bacterial membrane, while the second N-terminal domain, N2, may be responsible for binding of the bacterial F-pilus (reviewed in Spada et al., 1997). Recently, the structure of the N1-N2 domain complex has been solved at very high resolution (Lubkowski et al., 1998).

In phage display, a ligand (e.g. an antigen) is immobilized and a collection of binding proteins (e.g. antibodies) are displayed on the phage, that is, provided as fusions to the minor coat protein g3p (Winter et al., 1994). The essential trick is now that the genetic information of the displayed protein is contained within the phage DNA in the interior of the same phage particle (which is packed from the bacterial cell which makes the phage and produces the proteins) and thus, physically connected to the expressed protein. When the collection of antibody mutants is then passed over the antigen, only those which recognize the antigen will bind. They can then be eluted from the antigen and used to infect cells, and thereby the useful information, namely which sequence bound the antigen, can be amplified and deciphered.

This method is easy to understand when the protein variants differ in the binding site for the antigen. However, it also works when the collection of protein species differs in more global parameters, such as stability (Jung and Plückthun, 1997). The reason is that the experiment is carried out with populations of molecules: If a protein is unstable, only

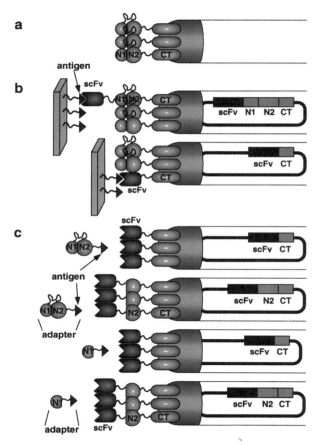

Figure 4. (a) On the top the w.t. phage is shown which contains g3p at one tip of the phage. The domain structure of g3p is indicated (for details, see text). (b) In conventional phage display, fusions of the protein of interest to the N-terminal domain (long fusions) or to the C-terminal domain (short fusion) are expressed together with a helper phage, which provides the g3p w.t., needed for infection. The desired phages are obtained from the mixture by "panning", that is binding to a solid surface, as schematically indicated. Each phage contains the DNA encoding the fusion protein. (c) SIP phages do not contain the N1 domain, which is absolutely needed for infection. N1 is provided by the adapter protein and must be linked to the phage by the cognate interaction. There are four different possibilities for constructing SIP phages (top to bottom), which differ in the phage either containing N2 or not, and in the adapter either containing N2 or not. The detailed consequences of these constructs are discussed elsewhere (Krebber et al., 1997).

few molecules from the population will be in the native state. If a phage displaying such a protein has to compete with one where a much larger percentage has reached the native state, the latter will bind much more frequently and outcompete the former. Thus, the selection is indirect and can use the "percentage of native molecules" as the ultimate parameter which is selected for.

The trouble with phage display is that it does not always work as smoothly, as the above description would suggest. The reason is that there are many ways how a phage can "stick" to the immobilized ligand: unfortunately, an unfolded protein or a partial sequence usually exposes a number of hydrophobic residues which stick strongly to the surface in question, but not because the molecules specifically recognizes the ligand (the antigen). Thus, several "rounds" are necessary, i.e., cycles of enrichment, in which the "right" binders are enriched over the "wrong" ones. However, the method is still not always successful. Therefore, two alternative methods, SIP and ribosome display, have been developed, described in the next two sections.

4.3. Selectively Infective Phage

The selectively infective phage (SIP) (Krebber et al., 1993; 1995; 1997; Duenas and Borrebaeck; 1994; Gramatikoff et al., 1994; Spada and Plückthun, 1997; Spada et al., 1997) is related to phage display technology using filamentous phages as described in the previous section. In contrast to phage display, the phage particles are rendered non-infective in SIP by disconnecting the N-terminal domains (N1 or N1-N2) of the phage g3p coat protein (Fig. 4b)–those involved in docking and bacterial cell penetration–from the C-terminal domain (CT) which caps the end of the phage. The N- and C-terminal domains are then each fused to one of the interacting partners being studied. One partner is thus displayed on the phage surface associated with CT, while the other is genetically fused or chemically coupled to the N-terminal domain(s), thus constituting a separate "adapter" molecule. Only when the specific protein-ligand interaction occurs between the partners is the g3p reconstituted and does the phage particle regain its infectivity, and the genetic information of a successful binder is propagated. In the absence of a cognate interaction, no infectivity is observed, demonstrating that the N-terminal domains really must be physically coupled to the phage for infection. In contrast, in phage display the whole phage binds and can be eluted from an immobilized target molecule, whereupon it can inject its DNA into bacterial cells and be amplified, since it remains infective.

SIP experiments can be carried out in vivo or in vitro (Fig. 5): in the former approach the phage or phagemid encodes both interacting molecules (e.g. antibody-antigen) in the same vector (Fig. 5a), and the protein-ligand interaction occurs in the bacterial periplasm; the infective phage particle is extruded from the cell and can be harvested and used to infect different bacteria. The simplicity of design of the experiment is counteracted to some extent by the lack of experimental control over parameters such as concentrations, incubation times and the infection process itself.

These factors can be controlled in the in vitro SIP experiment (Fig. 5b). In this case, the adapter protein is prepared from a separate E. coli culture, either by secretion or by in vitro refolding. This approach broadly extends the range of applications of SIP, as, in addition to genetically fused peptidic antigens, non-peptidic antigens can now be chemically coupled to the N-terminal domains. The adapter is then incubated with the non-infective phage library, and added to the bacteria.

A requirement of SIP appears to be a high affinity between protein and ligand, and all of the systems studied thus far have had dissociation constants of at least 10^{-8} M. It is

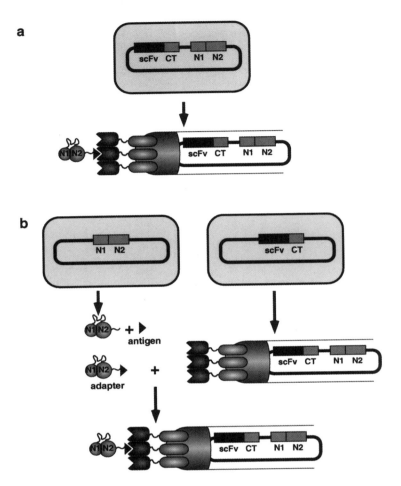

Figure 5. *In vitro* SIP vs. *in vivo* SIP. (**a**) The phage can encode both the ligand (fused to N1-N2 in this example) and the receptor (shown as a scFv fragment of an antibody in this example, fused to CT) (*in vivo* SIP). In this case, the *E. coli* cell produces an infective phage, if the ligand and receptor recognize each other, as the interaction already occurs in the bacterial periplasm upon budding of the phage. (**b**) Alternatively, a separate cell can be used to produce the adapter protein, which can be chemically coupled to the ligand (shown on the left), or provided as a fusion protein (not shown, analogous to (**a**)). The phage (as shown on the right) then does not encode the adapter and needs to be mixed *in vitro* with the adapter protein (*in vitro* SIP).

not yet clear, however, what the minimum affinity necessary for significant infectivity will be. Further studies will have to further quantitatively investigate the dependence of infectivity on properties of the proteins such as thermodynamic stability and expression rate.

No solid phase interaction with any support is necessary in SIP, limiting the potential of the occurrence of non-specific interactions and eliminating the need for elution. In addition, the low background infectivity observed with some of the SIP systems demonstrates that SIP can be an extremely effective and highly specific method of selecting for cognate interaction events. Furthermore, the exciting possibility of simultaneous screening of two interacting libraries by *in vivo* SIP might broaden the ability to identify interacting receptor-ligand pairs.

4.4. Ribosome Display

The development of the technology of "ribosome display" (Fig. 6) has been an effort to overcome a shortcoming all other technologies have in common, namely a limited library size (Fig. 7). The method was first developed for short peptides (Mattheakis et al., 1994), and then for whole proteins (Hanes and Plückthun, 1997). The trick is to carry out the selection entirely *in vitro*, without the use of any cells or phages at all. The concept is actually very simple. A pool of protein molecules is translated on ribosomes *in vitro* such that three conditions are fulfilled (Hanes and Plückthun, 1997). First, the protein must fold to is correct three-dimensional structure directly on the ribosome. Second, the protein must not leave the ribosome after the synthesis is finished. Third, the RNA (which carries the genetic information for the protein) must not leave the ribosome, either, after the synthesis is finished. Thereby, the phenotype and the genotype become linked *in vitro*—on the ribosome that makes the protein in the first place.

Before discussing the methods used to effect folding of the protein on the ribosome and preventing the dissociation of mRNA and protein, the advantages of this system shall be summarized. Since all steps occur *in vitro*, no transformation is necessary (Fig. 7). In all other selection systems, the DNA library needs to be taken up by bacteria. Usually this is done by electroporation (Dower and Cwirla, 1992), and while this method has been continuously optimized, only a tiny fraction of a typical ligation reaction is taken up by the cells. Typically, ligation reactions will be carried out with 10^{11}–10^{12} molecules of vector DNA, but in a single electroporation cuvette, typically only 10^7–10^8 molecules will be taken up to transform the bacteria—a tiny sample of the variety originally present. In the ribosome display technology, there is no transformation, and all the original library size is directly used in screening.

The ribosome display technology (Fig. 6) works by a modification of standard *in vitro* translation. A library of genes for protein molecules is amplified by PCR, and thereby a promoter is added at the 5' end. To provide resistance against nucleases, the mRNA also contains hairpins at both the 5' and the 3' end. Importantly, there is no stop codon present. The DNA library is transcribed, and the mRNA is isolated. This separate transcription and translation is advantageous as it allows different conditions for transcription and translation. Particularly, RNA polymerases require DTT, and the use of this compound during translation would largely prevent disulfide formation in the nascent protein. However, disulfides seem to be needed e.g. for the functional expression of antibody fragments (see also below). Therefore, translation is carried out separately from transcription, and the redox potential for the translation can be adjusted as needed and carried out in the presence of protein disulfide isomerase (PDI), which is essential for disulfide formation (Ryabova et al., 1997; Freedman et al., 1994). Furthermore, a whole series of combinations of molecular chaperones has been investigated and shown to further increase the yield of folded protein (Ryabova et al., 1997). The translation is then stopped by cooling on ice, and in the presence of very high Mg^{2+} concentrations, which stabilize the ribosomes against dissociation and thereby against release of mRNA and protein.

The method is made more efficient by also paying attention to the translation conditions. Vanadate transition state analog inhibitors of RNase are used. Additionally, an inhibitor of the *E. coli* 10Sa RNA (the product of the *ssrA* gene) is added. This RNA is specifically designed by the cell to remove proteins and mRNA which remain associated

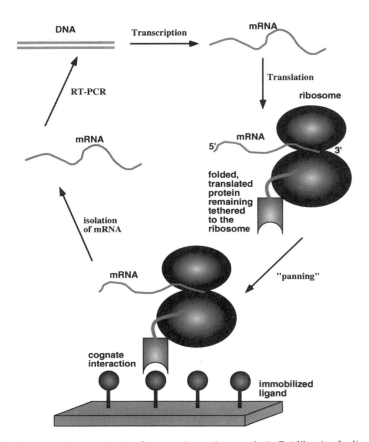

Figure 6. Principle of *in vitro* ribosome display for screening native protein (scFv) libraries for ligand (antigen) binding. A DNA scFv library is first amplified by PCR, whereby a T7 promoter, ribosome binding site and stem-loops are introduced, and then transcribed to RNA (top). After purification, mRNA is translated *in vitro* in an *E. coli* S-30 system in the presence of different factors enhancing the stability of ribosomal complexes and improving the folding of the scFv antibodies on the ribosomes (for details, see Hanes and Plückthun, 1997). Translation is then stopped by cooling on ice, and the ribosome complexes are stabilized by increasing the magnesium concentration. The desired ribosome complexes are affinity selected from the translation mixture by binding of the native scFv to the immobilized antigen (bottom). Unspecific ribosome complexes are removed by intensive washing. The bound ribosome complexes can then either be dissociated by EDTA or whole complexes can be specifically eluted with antigen. RNA is isolated from the complexes (left). Isolated mRNA is reverse transcribed to cDNA, and cDNA is then amplified by PCR. This DNA is then used for the next cycle of enrichment, and a portion can be analyzed by cloning and sequencing and/or by ELISA or RIA.

with the ribosome, when the mRNA does not have a stop codon (Keiler et al., 1996). However, since this association is the very basis of ribosome display, the 10Sa RNA must obviously be eliminated.

In studying the method first with a simple example of two antibodies, it was found that it is extremely efficient. The mRNAs were mixed at a ratio of $1:10^8$ and the dilute one was enriched over 5 cycles (Hanes and Plückthun, 1997). More exciting was the discovery that, by the use of *Taq* polymerase, which has an appreciable error rate of about 1 in 10^4 nucleotides (Keohavong and Thilly, 1989), mutations in the selected molecules were

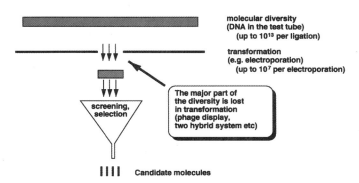

Figure 7. Schematic representation of the transformation problem. In the creation of a library *in vitro*, a large number of diverse molecules is created and—usually—eventually ligated to a plasmid. Upon transformation of cells (usually, electroporation of *E. coli*), only a small fraction of this library is actually taken up by the cells and subjected to phage production and panning. Thus, only a small fraction of the library actually takes part in the selection process. This problem can be alleviated by repeating the electroporation process many times, until the desired library size is achieved, or by using cell-free panning strategies.

found: the population has clearly evolved to remain compatible with binding. It is this easy introduction of mutations which is a most exciting prospect, as it may allow us to increase the stringency of selection and thereby truly improve a phenotype in real time. Since all steps occur *in vitro*, mutagenesis and selection is much faster achieved than in the phage based methods detailed above. Nevertheless, some proteins are more easily enriched than others, and the molecular reasons for this are still being investigated.

5. CASE STUDY I: A STABLE DISULFIDE-FREE ANTIBODY SCFV FRAGMENT BY DIRECTED EVOLUTION

Evolutionary technology has been applied to a number of different problems in protein optimization (summarized in the Introduction). Here, the power of this approach for studying problems of protein structure and function shall be exemplified with two studies on antibody structure.

In the first study, an evolutionary approach was used to generate a disulfide free antibody single-chain Fv (scFv) fragment (Proba et al., 1998). In a scFv fragment, the variable domain of the heavy chain (V_H) is connected by a flexible linker to the variable domain of the light chain (V_L) (reviewed by Huston et al., 1993). The Fv or scFv fragment is thus the smallest unit still containing the full binding site of the antibody.

The antibody domain consists of a beta sandwich, with five and four strands making up the two sheets (Fig. 2). There is a very conserved disulfide bond, connecting strands b and f. About 99.5% of all sequences (murine and human) in the Kabat database and all the germline genes of the human antibodies (Vbase, version 1997, http://www.mrc-cpe.cam.ac.uk/imt-doc/public/INTRO.html) contain both of the cysteines making up this disulfide bond.

In the cytoplasm of the cell, the redox potential is reducing, and no disulfide bonds can form (Gilbert, 1990). Thus, when an antibody is expressed in the cytoplasm of a cell, its disulfide bonds do not form, and most frequently the native state is not stable. There are many pieces of evidence that the disulfide bond is critical for stability. First, refolding experiments in the presence of reducing agents and directed mutagenesis experiments have been used to replace the cysteines against other hydrophobic residues in many combinations. The stability is severely affected and sometimes, a stable protein is not obtained at all (Goto and Hamaguchi, 1979; Glockshuber et al., 1992). Nevertheless, a few natural antibodies exist (Rudikoff and Pumphrey, 1986) which are missing one of the cysteines, which must have been lost by somatic mutations occurring in the maturation of the B-cell. Since at least a few of these antibodies have been found to be functional, in these cases the stable structure seems to be reached anyway, even in the absence of the stabilizing disulfide bond. It appears, therefore, that additional stabilizing mutations can compensate for the loss of the contribution of the disulfide bond to stability. Indeed, in a study of a V_L domain, stabilizing mutations were shown to be able to partially compensate for the loss of the disulfide and lead to a protein of reduced stability (Frisch et al., 1994, 1996), but normal structure (Uson et al., 1997).

For a number of applications, it would be of great interest to construct antibodies which do not need the disulfide bond for stability, and can be expressed inside the cell as "intrabodies" (Richardson and Marasco, 1995; Biocca and Cattaneo, 1995). For example, in the various genome projects, an enormous number of new proteins will be described for which no function will be known, and it would be very helpful to establish a function by titrating out the protein in question and observing the effect is has on cellular functions. Since a recombinant antibody against the recombinant protein can be obtained very quickly (Winter et al., 1994), it would be very attractive to use such an antibody for this purpose. Thus, in many cases, the antibody would have to be functional in the cytoplasm. More ambitiously, the antibody might be a catalytic one (Lerner et al., 1991), and a functional cytoplasmic expression might be useful for integrating the new reaction into general metabolism to either select for higher activities or even, in the very long run, redirect metabolism into new directions to produce new metabolites and new pharmacophores in plants or fungi. Furthermore, cytoplasmic and nuclear expression may also be used to couple antibody antigen interaction to transcription, to either obtain a readout system by generating a color or light as a consequence of the cognate interaction or to turn on a selectable gene as a consequence. Thus, there is clearly sufficient motivation to generate functional disulfide-free scFv fragments.

Directed evolution was used to achieve this goal (Proba et al., 1998). The strategy was to select for stabilizing mutations anywhere in the structure which might overcome the loss of the disulfide bond (Fig. 8). As a starting protein, an antibody was used in which the disulfide bond in the heavy chain was missing already. In this case, Cys H92 is replaced by Tyr in the natural protein, and this protein had been shown before to be functional (Rudikoff and Pumphrey, 1986). In order to "gently" apply the evolutionary pressure, the two domains V_H and V_L were first independently optimized, then combined under further mutagenesis. Thus, the best combination of cysteine-free domains was selected and further mutations were introduced to compensate for the loss of the disulfide.

In a first set of experiments, the missing cysteine was restored in the heavy chain, such that this domain carried a disulfide bond, and in the light chain, the positions of the cysteines were randomized to Ala, Val, Leu, Ile and Phe (Fig. 8). In a parallel set of experiments, the position H22 in the V_H domain was randomized to Ala, Val, Leu, and Ser, position H98 already being tyrosine. Both libraries were screened by classical phage display, and only the combinations in Fig. 8 (third line) survived the selection.

The two sets of disulfide free domains were then combined to identify proteins which were functional in the absence of any cysteines in any domain. However, the selected domains were not simply ligated, but a DNA shuffling (Stemmer, 1994a) was carried out. This

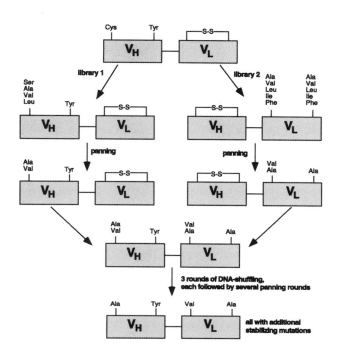

Figure 8. Schematic description of the directed evolution of disulfide-free scFv fragments. As a model system, the scFv fragment ABPC48 (naturally missing the disulfide in V_H, top line) was used. Two libraries were constructed, each restoring the disulfide in one domain and randomizing the cysteine positions in the other. The selected variants were recombined and subjected to DNA shuffling (Fig. 3c), and the final version, not containing any cysteines, was only found in the presence of additional mutations, but is practically unable to fold in the absence of further stabilizing mutations. (For details, see Proba et al., 1998).

was done to introduce additional mutations which may stabilize the molecules. Over several rounds of shuffling and selection, a range of new variants were obtained (Proba et al., 1998). The conditions for screening were slowly increased in stringency.

The result of the directed evolution experiment was a series of molecules, able to bind the antigen and not containing any cysteines. Importantly, the control molecule which only contained the cysteine replacements and not any of the additional mutations could not even be produced to test its stability (Proba et al., 1998). A number of different mutations were thus selected which all improve stability, but only by small increments. However, the combination of these mutations in the same molecule, which is one of the hallmarks of the evolutionary approach, solved the problem of making a stable, disulfide-free scFv fragment.

Some of the mutations can now be rationalized. For instance, the mutation LysH66Arg (Fig. 9) can be modeled, since this residue occurs naturally in human antibodies. It forms an ideal interaction with Asp-H86, with the two guanidino NH-groups of Arg forming hydrogen bonds to the two carboxylate oxygens of Asp. This salt bridge is quite buried, increasing its stabilizing effect compared to that of a solvent-exposed ion pair. The elucidation of the exact mechanism of how other mutations improve the activity of the protein clearly requires further studies, such as e.g. for the stabilizing mutation AsnH52Ser (Fig. 9).

Other mutations may have been selected for the efficiency of folding, and not thermodynamic stability. The ability of antibody mutants to avoid aggregation appears to be not mechanistically linked to thermodynamic stability (Knappik and Plückthun, 1995; Nieba et al., 1997; Jung and Plückthun, 1997), and it is possible that some mutants of this type have been selected.

In conclusion, it can be seen that the evolutionary approach selected a series of very subtle improvements which would have been extraordinarily hard to predict, but are now ideal objects for further studies. Thus, while the same amount of physical chemistry needs to be carried out to increase the understanding of the effect of the mutations as would have been necessary in a site directed mutagenesis approach, it is done with a much more relevant set of mutants, namely those which all lie on a path to higher stability and a higher yield of folding.

6. CASE STUDY II: THE KINK IN THE FIRST STRAND OF ANTIBODY LIGHT CHAINS

Another example which was studied with directed molecular evolution, this time using SIP (see above), is the optimal configuration of the first strand of the β-sandwich. A

a   Lys H66 ->Arg

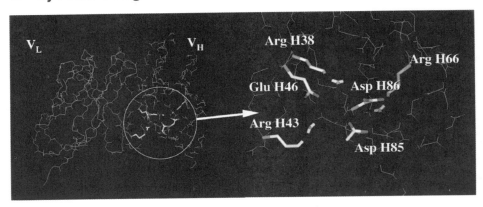

b   Asn H52 -> Ser

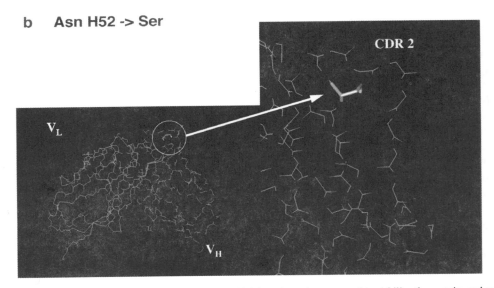

Figure 9. Two of the selected mutations are shown, which have been demonstrated to stabilize the protein, and are essential for the disulfide-free protein to remain stable. (For details, see Proba et al., 1998).

hallmark of the immunoglobulin variable domain is the first β-strand, which is interrupted. The first part of the strand makes hydrogen bonds, typical of an antiparallel arrangement, to the outer β-sheet of the V_L/V_H heterodimer. Next follows a kink, whose presence but not sequence is conserved in antibody variable domains, and then another strand which makes hydrogen bonds to the inner β-sheet of the domain. Thus, this kink is essential in closing off the sides of the sandwich to solvent, and it thus helps to connect the two sheets (Fig. 10).

Nature has found several solutions to this problem. In V_κ domains a *cis*-proline in position L8 (sequence numbers according to Kabat et al., 1991) is strongly conserved. In V_λ domains, the kink is one amino acid shorter (structurally, amino acid L7 is missing), and usually contains at least one *trans*-proline (Figure 10). V_H domains, which are similar to V_λ in length, contain glycine residues in positions H8 and/or H10, frequently with positive Phi angles, to make the kink (A. Honegger, unpublished).

It was therefore of interest to investigate which of these solutions to the kink would be selected in a directed evolution experiment. A scFv fragment (specific for a hemagglutinin peptide, PDB file 1ifh) was chosen as a model system, and it contains a kappa light chain. Two small libraries were constructed in which either residues L7, L8 and L9 were randomized, or replaced by only two random amino acids. Thus two libraries were obtained, one of the length of kappa chains, and one of the length of lambda chains at the kink.

Several rounds of SIP selection were carried out. In the first series of experiments, the "lambda length" variants were kept separate from the kappa length variants. After only one selection round of the "kappa length", 50% of all selected molecules already carried the *cis*-proline typical for these molecules. After a second round, all molecules carried the *cis*-proline. After the third round, the flanking residues began to take shape and very similar preferences were selected as found in the Kabat database of antibody sequences (Fig. 10) (S. Spada, A. Honegger and A. Plückthun, manuscript submitted).

Similarly, in the "lambda length" variants, after only one round, 50% proline were found for either position, which increased to 100% of the molecule having either one of the two of the random positions being proline, with a significant portion having both residues as proline. Interestingly, again these are the variants found by far most frequently in the Kabat database. In a third set of experiments, the "kappa-length" library and the "lambda-length" library were mixed. The kappa sequences were indeed selected, and again a distribution resulted with the most common sequences of kappa domains enriched (S. Spada, A. Honegger and A. Plückthun, manuscript submitted).

Thus, directed molecular evolution reproduces quite faithfully the evolution of the natural molecules, and thus the nature of the selective pressure must have been similar. The molecular parameters why these variants were selected was investigated. The selected variants were found to be more stable against urea denaturation (S. Spada, A. Honegger and A. Plückthun, manuscript submitted). This shows again that complex molecular parameters such as stability can be optimized by molecular evolution.

7. CONCLUSIONS AND PERSPECTIVES

The strategy of using directed molecular evolution is a new tool in the study of proteins. It allows the investigator to concentrate on those mutants which are on the pathway to improved function. As it is quite independent of the availability of working hypotheses, it may possibly uncover new relationships in the protein structure. Clearly, such work does

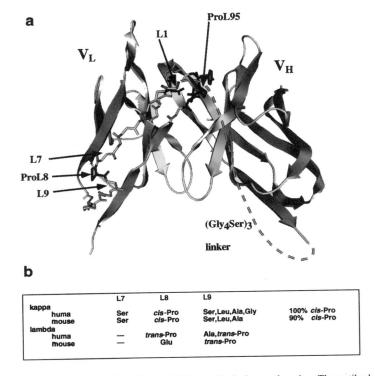

Figure 10. (a) Location of the cis-prolines L8 and L95 in antibody kappa domains. The antibody is represented according to secondary structure, except in the CDR3 region, and the N-terminal part of the kappa domain, where the backbone is shown. This serves to indicate where the randomized region (L7-L9) is situated with respect to the whole protein. **(b)** The predominant sequences found in kappa and lambda light chains at positions L7, L8 and L9. Lambda chains are one amino acid shorter, and superpositions suggests to make the deletion at position L7.

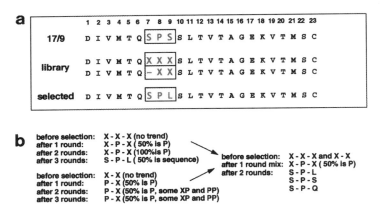

Figure 11. (a) The sequence of the starting antibody 17/9 in the N-terminal region of V_L is shown. Two libraries were constructed, containing the same length as in kappa domains (denoted XXX) and one amino acid shorter and thus corresponding to lambda domains (denoted–XX). The most frequently selected molecule has the sequence shown on the bottom. **(b)** The results of the SIP experiment are shown in more detail, as are the predominant features obtained after different numbers of rounds from the "XXX" library and the "XX" library (left), and after combination of the libraries (right). It can be seen that the predominant sequences found in the Kabat library of antibody sequences are remarkably well selected.

not obviate any biophysical investigations whatsoever. On the contrary, since the sequence space is so enormous, the evolutionary approach often still needs to be focused, as a complete sampling is impossible. Thus, structural and evolutionary studies are likely to flourish best when used in combination. The main challenge for the future is to truly "direct" evolution, that is to learn how to apply the pressure exclusively to the molecular quantity of interest. It appears that directed molecular evolution is ultimately a very "rational" approach.

ACKNOWLEDGMENTS

I am indebted to all my coworkers mentioned in the text and references, who have developed the technologies described in this article and have been a constant source for stimulating discussions. Particularly, Dr. Annemarie Honegger has been instrumental in the molecular modeling for the projects described and an invaluable intellectual contributor to all structural issues. I thank Jozef Hanes, Frank Hennecke, Annemarie Honegger, Lutz Jermutus, and Peter Lindner for critically reading the manuscript.

REFERENCES

Baca, M., Scanlan, T. S., Stephenson, R. C. and Wells, J. A. (1997). Phage display of a catalytic antibody to optimize affinity for transition-state analog binding. *Proc. Natl. Acad. Sci. USA* **94**, 10063–10068.

Bai, C. and Elledge, S. J. (1996). Gene identification using the yeast two-hybrid system. *Methods Enzymol.* **273**, 331–347.

Boder, E. T. and Wittrup, K. D. (1997). Yeast surface display for screening combinatorial polypeptide libraries. *Nature Biotechnol.* **15**, 553–557.

Biocca, S. and Cattaneo, A. (1995). Intracellular immunization: Antibody targeting to subcellular compartments. *Trends Cell Biol.* **5**, 248–252.

Cadwell, R. C. and Joyce, G. F. (1994). Mutagenic PCR. *PCR Methods Appl.* **3**, 136–140.

Cortese, R., Monaci, P., Luzzago, A., Santini, C., Bartoli, F., Cortese, I., Fortugno, P., Galfre, G., Nicosia, A. and Felici, F. (1996). Selection of biologically active peptides by phage display of random peptide libraries. *Curr. Opin. Biotechnol.* **7**, 616–621.

Cull, M. G., Miller, J. F. and Schatz, P. J. (1992). Screening for receptor ligands using large libraries of peptides linked to the C terminus of the lac repressor. *Proc. Natl. Acad. Sci. USA* **89**, 1865–1869.

Dower, W. J. and Cwirla, S. E. (1992) in *Guide to Electroporation and Electrofusion*, Eds.: Chang, D. C., Chassy, B. M., Saunders, J. A. and Sowers, A. E., Academic Press, San Diego, 291–301.

Duenas, M. and Borrebaeck, C. A. (1994). Clonal selection and amplification of phage displayed antibodies by linking antigen recognition and phage replication. *Bio/Technol.* **12**, 999–1002.

Dunn, I. S. (1996). Phage display of proteins. *Curr. Opin. Biotechnol.* **7**, 547–553.

Fersht, A. R. (1995). Characterizing transition states in protein folding: an essential step in the puzzle. *Curr. Opin. Struct. Biol.* **5**, 79–84.

Fersht, A. R. (1997). Nucleation mechanisms in protein folding. *Curr. Opin. Struct. Biol.* **7**, 3–9.

Fersht, A. R. and Wells, T. N. (1991). Linear free energy relationships in enzyme binding interactions studied by protein engineering. *Protein Eng.* **4**, 229–231.

Freedman, R. B., Hirst, T. R. and Tuite, M. F. (1994). Protein disulphide isomerase: building bridges in protein folding. *Trends Biochem. Sci.* **19**, 331–336.

Frisch, C., Kolmar, H. and Fritz, H. J. (1994). A soluble immunoglobulin variable domain without a disulfide bridge: Construction, accumulation in the cytoplasm of E. coli, purification and physicochemical characterization. *Biol. Chem. Hoppe-Seyler* **375**, 353–356.

Frisch, C., Kolmar, H., Schmidt, A., Kleemann, G., Reinhardt, A., Pohl, E., Usón, I., Schneider, T. R. and Fritz H. J. (1996). Contribution of the intramolecular disulfide bridge to the folding stability of REI(v), the variable domain of a human immunoglobulin kappa light chain. *Fold. Des.* **1**, 431–440.

Georgiou, G., Stathopoulos, C., Daugherty, P. S., Nayak, A. R., Iverson, B. L. and Curtiss, R,III (1997). *Nature Biotechnol.* **15**, 29–34.

Gilbert, H. F. (1990). Molecular and cellular aspects of thiol-disulfide exchange. *Adv. Enzymol.* **63**, 69–172.

Glockshuber, R., Schmidt, T. and Plückthun, A. (1992). The disulfide bonds in antibody variable domains: effects on stability, folding *in vitro*, and functional expression in *Escherichia coli*. *Biochemistry* **31**, 1270–1279.

Goto, Y. and Hamaguchi, K. (1979). The role of the intrachain disulfide bond in the conformation and stability of the constant fragment of the immunoglobulin light chain. *J. Biochem.* **86**, 1433–1441.

Gramatikoff, K., Georgiev, O. and Schaffner, W. (1994). Direct interaction rescue, a novel filamentous phage technique to study protein-protein interactions. *Nucleic Acids Res.* **22**, 5761–5762.

Hall, A. and Knowles, J. R. (1976). Directed selective pressure on a beta-lactamase to analyse molecular changes involved in development of enzyme function. *Nature* **264**, 803–804.

Hanes, J. and Plückthun, A. (1997). *In vitro* selection and evolution of functional proteins using ribosome display. *Proc. Natl. Acad. Sci. USA* **94**, 4937–4942.

Horovitz, A. (1996). Double-mutant cycles: a powerful tool for analyzing protein structure and function. *Fold. Des.* **1**, 121–126.

Huston, J. S., McCartney, J., Tai, M. S., Mottola-Hartshorn, C., Jin, D., Warren, F., Keck, P. and Oppermann, H. (1993). Medical applications of single-chain antibodies. *Intern. Rev. Immunol.* **10**, 195–217.

Jin, L., Fendly, B. M. and Wells, J. A. (1992). High resolution functional analysis of antibody-antigen interactions. *J. Mol. Biol.* **226**, 851–865.

Jung, S. and Plückthun, A. (1997). Improving *in vivo* folding and stability of a single-chain Fv antibody fragment by loop grafting. *Prot. Eng.* **10**, 959–966.

Kabat, E. A., Wu, T. T., Perry, H. M., Gottesman, K. S. and Foeller, C. (1991). *Sequences of Proteins of Immunological Interest*, 5th Ed., National Institutes of Health, Bethesda, MD.

Kayushin, A. L., Korosteleva, M. D., Miroshnikov, A. I., Kosch, W., Zubov, D. and Piel, N. (1996). A convenient approach to the synthesis of trinucleotide phosphoramidites--synthons for the generation of oligonucleotide/peptide libraries. *Nucleic Acids Res.* **24**, 3748–3755.

Keiler, K. C., Waller, P. R. and Sauer, R. T. (1996). Role of a peptide tagging system in degradation of proteins synthesized from damaged messenger RNA. *Science* **271**, 990–993.

Keohavong, P. and Thilly, W. G. (1989). Fidelity of DNA polymerases in DNA amplification. *Proc. Natl. Acad. Sci. USA* **86**, 9253–9257.

Knappik, A. and Plückthun, A. (1995). Engineered turns of an antibody improve its *in vivo* folding. *Protein Eng.* **8**, 81–89.

Krebber, C., Moroney, S., Plückthun, A. and Schneider, C. (1993). European Patent Application EP 93102484.

Krebber, C., Spada, S., Desplancq, D. and Plückthun, A. (1995). Coselection of cognate antibody-antigen pairs by selectively-infective phages. *FEBS Lett.* **377**, 227–231.

Krebber, C., Spada, S., Desplancq, D., Krebber, A., Ge, L. and Plückthun, A. (1997). Selectively infective phage (SIP): a mechanistic dissection of a novel method to select for protein-ligand interactions. *J. Mol. Biol.* **268**, 619–630.

Lerner, R. A., Benkovic, S. J. and Schultz, P. G. (1991). At the crossroads of chemistry and immunology: catalytic antibodies. *Science* **252**, 659–667.

Low, N. M., Holliger, P. H. and Winter, G. (1996). Mimicking somatic hypermutation: affinity maturation of antibodies displayed on bacteriophage using a bacterial mutator strain. *J. Mol. Biol.* **260**, 359–368.

Lubkowski, J., Hennecke, F., Plückthun, A. and Wlodawer A. (1998). The structural basis of phage display: The crystal structure of the N-terminal domains of g3p at 1.46 Å resolution. *Nature Structural Biology*, **5**, 140–147.

Mattheakis, L. C., Bhatt, R. R. and Dower, W. J. (1994). An *in vitro* polysome display system for identifying ligands from very large peptide libraries. *Proc. Natl. Acad. Sci. USA* **91**, 9022–9026.

Moore, J. C., Jin, H. M., Kuchner, O. and Arnold, F. H. (1997). Strategies for the *in vitro* evolution of protein function: enzyme evolution by random recombination of improved sequences. *J. Mol. Biol.* **272**, 336–347.

Nieba, L., Honegger, A., Krebber C. and Plückthun, A. (1997). Disrupting the hydrophobic patches at the antibody variable/constant domain interface: improved *in vivo* folding and physical characterization of an engineered scFv fragment. *Protein Eng.* **10**, 435–444.

Padlan, E. A. (1996). X-ray crystallography of antibodies. *Adv. Protein Chem.* **49**, 57–133.

Phizicky, E. M. and Fields, S. (1995). Protein-protein interactions: methods for detection and analysis. *Microbiol. Rev.* **59**, 94–123.

Proba, K., Wörn, A., Honegger, A. and Plückthun, A. (1998). Antibody fragments without disulfide bonds, made by molecular evolution. *J. Mol. Biol.* **275**, 245–253.

Richardson, J. H. and Marasco, W. A. (1995). Intracellular antibodies: development and therapeutic potential. *Trends Biotechnol.* **13**, 306–310.

Rudikoff, S. and Pumphrey, J. G. (1986). Functional antibody lacking a variable-region disulfide bridge. *Proc. Natl. Acad. Sci. USA* **83**, 7875–7878.

Ryabova, L., Desplancq, D., Spirin, S. and Plückthun, A. (1997). Making antibodies *in vitro*: the effect of molecular chaperones and disulfide isomerase on the functional expression of single-chain fragments by *in vitro* translation. *Nat. Biotechnol.***15**, 79–84.

Scott, J. K. and Smith, G. P. (1990). Searching for peptide ligands with an epitope library. *Science* **249**, 386–390.

Smith, G. P. (1985). Filamentous fusion phage: novel expression vectors that display cloned antigens on the virion surface. *Science* **228**, 1315–1317.

Spada, S. and Plückthun, A. (1997). Selectively infective phage (SIP) technology: a novel method for the *in vivo* selection of interacting protein-ligand pairs. *Nature Med.* **6**, 694–696.

Spada, S., Krebber, C. and Plückthun, A. (1997). Selectively infective phages. *Biol. Chem.* **378**, 445–456.

Stemmer, W. P. (1994a). Rapid evolution of a protein *in vitro* by DNA shuffling. *Nature* **370**, 389–391.

Stemmer, W. P. (1994b). DNA shuffling by random fragmentation and reassembly: *in vitro* recombination for molecular evolution. *Proc. Natl. Acad. Sci. USA* **91**, 10747–10751.

Usón, I., Bes, M. T., Sheldrick, G. M., Schneider, T. R., Hartsch, T. and Fritz, H. J. (1997). X-ray crystallography reveals stringent conservation of protein fold after removal of the only disulfide bridge from a stabilized immunoglobulin variable domain. *Fold. Des.* **2**, 357–361.

Virnekäs, B., Ge, L., Plückthun, A., Schneider, K. C., Wellnhofer, G. and Moroney, S. E. (1994). Trinucleotide phosphoramidites: ideal reagents for the synthesis of mixed oligonucleotides for random mutagenesis. *Nucl. Acids Res.* **22**, 5600–5607.

Wagner, S. D. and Neuberger, M. S. (1996). Somatic hypermutation of immunoglobulin genes. *Annu. Rev. Immunol.* **14**, 441–457.

Warbrick, E. (1997). Two's company, three's crowd: the yeast two hybrid system for mapping molecular interactions. *Structure* **5**, 13–17.

Ward, W. H., Timms, D. and Fersht, A. R. (1990). Protein engineering and the study of structure-function relationships in receptors. *Trends Pharmacol. Sci.* **11**, 280–284.

Wells, J. A., Cunningham, B. C., Fuh, G., Lowman, H. B., Bass, S. H., Mulkerrin, M. G., Ultsch, M. and deVos, A. M. (1993). The molecular basis for growth hormone-receptor interactions. *Recent Prog. Horm. Res.* **48**, 253–275.

Winter, G., Griffiths, A. D., Hawkins, R. E. and Hoogenboom, H. R. (1994). Making antibodies by phage display technology. *Annu. Rev. Immunol.* **12**, 433–455.

5

HIGH PRESSURE EFFECTS ON PROTEIN STRUCTURE*

Kenneth E. Prehoda,† Ed S. Mooberry, and John L. Markley

Department of Biochemistry
University of Wisconsin-Madison
420 Henry Hall
Madison, Wisconsin 53706

1. INTRODUCTION

The protein stability problem is one of the most intriguing questions in protein chemistry. The central conundrum is why the folded (native) form of a protein, with its specific three-dimensional structure, is stabilized relative to the unfolded (denatured) form, which is conformationally disperse. The related problems of protein dynamics, stability, and design are all related ultimately to protein energetics. Our knowledge of the *microscopic* characteristics of proteins is constantly growing as more three-dimensional structures are determined by X-ray crystallography and nuclear magnetic resonance spectroscopy. To fully understand the energetic details of proteins, it is necessary to have information not only about their structures but also of their *macroscopic* behavior. A fundamental approach to protein energetics is the study of thermodynamics, i.e., the equilibrium position of the system (Callen, 1985). In the context of protein stability, this equilibrium is between the active, folded form, of a protein and the inactive, conformationally disperse, unfolded form. Not surprisingly, under physiological conditions this equilibrium usually favors the native form. By examining the response of this equilibrium process to different perturbants, we can infer a remarkable amount of information regarding the energetics of protein folding. In this chapter, we focus on the use of pressure as a variable for perturbing equilibria connecting protein structural states. As detailed below, the standard

* This work was supported by NIH grant GM35976. High pressure probe development and NMR spectroscopy were carried out at the National Magnetic Resonance Facility at Madison which is funded by NIH grant RR02301; equipment in the Facility was purchased with funds from the University of Wisconsin, the NSF Biological Instrumentation Program (grant DMB-8415048), the National Biomedical Research Technology Program (grant RR02301), NIH Shared Instrumentation Program (grant RR02781), and the U.S. Department of Agriculture

† Present address: Department of Cellular and Molecular Pharmacology, University of California, San Francisco, 513 Parnassus Ave., San Francisco California 94143

Protein Dynamics, Function, and Design, edited by Jardetzky *et al.*
Plenum Press, New York, 1998

volume change for a reaction is the central component of pressure studies, although higher order terms, such as the standard compressibility change can also be important. Volume changes associated with reactions can be determined from investigations of the pressure dependence of equilibria; activation volumes for reactions can be obtained from measurements of pressure effects on reaction rates (Markley et al., 1996).

1.1. Early Studies of Pressure Effects on Proteins

The earliest report of pressure effects on proteins dates back to the early part of the 20th century when Bridgman (1914) observed that elevated hydrostatic pressure irreversibly denatures egg albumin. The late 1960's and early 1970's saw the first extensive high pressure studies of proteins. In particular, reversible pressure denaturation was demonstrated for ribonuclease A (Brandts et al., 1970; Gill & Glogovsky, 1965), chymotrypsinogen (Hawley, 1971), and metmyoglobin (Zipp & Kauzmann, 1973). This list was extended recently to include lysozyme (Samarasinghe et al., 1992) and staphylococcal nuclease (Royer et al., 1993). The observed unfolding volumes found in these studies are typically between -15 and -100 mL mol^{-1} at 298K and atmospheric pressure (Table I). These changes are small, considering that the partial molar volumes of these proteins are greater than 10,000 mL mol^{-1}. The magnitude and sign of the experimental ΔV° values indicate that the partial molar volumes of the folded and unfolded states are similar, with the unfolded state smaller by approximately 0.5%. An underlying theme of these studies has been the lack of a consistent explanation for the physical basis of the small and negative volume change for denaturation. This recently has been termed the "protein volume paradox" (Chalikian & Breslauer, 1996).

A number of studies have demonstrated that, in general, oligomeric proteins are dissociated at pressures lower than those required to unfold the individual subunits of the complex. Reaction volumes for oligomer dissociation reactions taken from the literature are shown in Table II.

1.2. Thermodynamics of the Pressure Denaturation Process

The first and second laws of thermodynamics describe how the internal energy E of a system changes as reversible work is performed on the system, or as heat is exchanged with the surroundings,

$$dE = TdS - pdV + dw'_{rev} \tag{1}$$

Table I. Protein Denaturation Reaction Volumes

Protein	ΔV°_{obs} (/mL mol^{-1})
Ribonuclease A	$-6^a, -46^b, -56^c$
Staphylococcal nuclease	-85^d
Chymotrypsinogen	-14^e
Lysozyme	-26^f
Metmyoglobin	-60^g
Flavodoxins	-64^h
FMN-binding protein	-74^i

^{a}(K. E. Prehoda, E. S. Mooberry, & J. L. Markley, manuscript in preparation); ^{b}(Brandts et al., 1970); ^{c}(Zhang et al., 1995); ^{d}(Royer et al., 1993); ^{e}(Hawley, 1971); ^{f}(Li et al., 1976); ^{g}(Zipp & Kauzmann, 1973); ^{h}(Visser et al., 1977); ^{i}(Li et al., 1976).

Table II. Examples of Pressure Induced Dissociation
of Dimeric and Tetrameric Proteins

Protein	ΔV°_{obs} (/mL mol^{-1})
Dimers	
Enolase	-55^{a}
Hexokinase	-120^{b}
Arc repressor	-100^{c}
Rubisco	-130^{d}
Tryptophan synthase B2 subunit	-170^{e}
Hemoglobin dimer	-90^{f}
R17 coat protein	-130^{g}
Cyt b_{5}-Cyt c	-122^{h}
Tetramers	
Lactate dehydrogenase	-170^{i}
GADPH	-235^{j}
Lac repressor	-170^{k}
Phosphorylase a	-200^{l}

[a](Kornblatt et al., 1993), [b](Ruan & Weber, 1988), [c](Oliveira et al., 1994), [d](Erijman et al., 1993), [e](Silva et al., 1986), [f](Pin et al., 1990), [g](De Poian et al., 1993), [h](Rodgers et al., 1988), [i](King & Weber, 1986), [j](Ruan & Weber, 1989), [k](Royer et al., 1986), [l](Ruan & Weber, 1993).

where p is the pressure; V the volume; T the absolute temperature; S the entropy; and w'_{rev} represents non-pV work. As we are concerned with the equilibrium position of the system at constant temperature and pressure, the Gibbs potential, G, is a particularly useful variable:

$$dG \equiv dE + pdV + Vdp - TdS - SdT \tag{2}$$

Upon writing the external reversible work in the explicit form, $dw'_{rev} = XdY$, which represents some form of non pV work, and combining equations 1 and 2, one obtains

$$dG = -SdT + Vdp + XdY \tag{3}$$

Comparison of Eq. 3 to the general expression for the differential of $G(T,p,Y)$

$$dG = \left(\frac{\partial G}{\partial T}\right)_{p,Y} dT + \left(\frac{\partial G}{\partial p}\right)_{T,Y} dp + \left(\frac{\partial G}{\partial Y}\right)_{p,T} dY \tag{4}$$

allows us to equate the pressure derivative of the Gibbs free energy to the volume:

$$\left(\frac{\partial G}{\partial p}\right)_{T,Y} = V \tag{5}$$

An important aspect of protein energetics is the interaction between the protein and its solvent. If the solution contains only two components (1 = solvent, 2 = solute), the chemical potential of the protein μ_2 is given by

$$\mu_2 \equiv \left(\frac{\partial G}{\partial n_2}\right)_{T,p,n_1}$$

(6)

The standard free energy change for a reaction is given by the difference in standard chemical potentials of the final (*f*) and initial (*i*) states:

$$\Delta \overline{G}^\circ = \mu_f{}^\circ - \mu_i{}^\circ$$

(7)

In Eq. 7, $\mu_k{}^\circ$ represents a hypothetical standard state in which each species *k* (where *k* = *i* or *f*) is at unit concentration, but all interactions are those of the dilute solution. The pressure dependence of the reaction is then,

$$\left(\frac{\partial \Delta \overline{G}^\circ}{\partial P}\right)_T = \left(\frac{\partial \mu_f{}^\circ}{\partial P}\right)_T - \left(\frac{\partial \mu_i{}^\circ}{\partial P}\right)_T = \overline{V}^\circ{}_f - \overline{V}^\circ{}_i = \Delta \overline{V}^\circ$$

(8)

where $\overline{V}^\circ{}_k$ is the standard-state partial molar volume of species *k*. The standard molar free energy change of a reaction is related to its equilibrium constant by the familiar expression

$$\Delta \overline{G}^\circ = -RT \ln K_{eq}$$

(9)

For an intuitive understanding of the effect of pressure on a reaction, it is useful to recall the principle of Le Châtelier, which can be stated in relation to pressure effects (Hamann, 1980) as, "If a thing can shrink, it *will* shrink if you squeeze it." In other words, the effect of pressure on a reaction is to move it in the direction of smaller volume. If products and reactants have equal contributions to the volume of the mixture, then pressure will not change the equilibrium position of the system. Since proteins denature at high pressure, this implies that the partial molar volume of the unfolded protein is less than that of the folded protein.

2. HIGH PRESSURE METHODOLOGY

In comparison to the numerous studies of the effects on proteins of temperature and chemical denaturants, the literature contains relatively few reports of pressure effects on proteins. Technical difficulties involved in containing the very high pressures required to perturb protein equilibria have contributed to this discrepancy. Quantitative denaturation of a fairly stable protein may require pressures in excess of 10 kbar. To illustrate the magnitude of these pressures, consider that 1 bar = 0.987 atm = 9.93 meter sea water. At the deepest point in the ocean, the Mariana Trench (11,021 meter), the pressure generated by the depth of sea water is only about 1.1 kbar. Containing pressures this great in an experimental apparatus, while still being able to observe the desired reaction, represents a significant technical challenge (Bridgman, 1931; Holzapfel & Issacs, 1997). By comparison, achieving the elevated temperatures necessary to study protein denaturation (often less than 70°C) is a simple proposition.

2.1. Detection of Pressure-Induced Changes in Proteins

The most prevalent approaches to monitoring pressure-induced changes in protein structure involve branches of optical spectroscopy such as absorption, fluorescence, or circular dichroism at ultraviolet, visible, or infrared frequencies. These techniques typically exploit the dependence of signals from intrinsic chromophores on the state of the protein to evaluate the extent of the reaction.

2.1.1. Optical Spectroscopy at High Pressure. A high pressure vessel for use with optical spectroscopies has been described (Drickamer et al., 1972). The system consists of a large steel cylinder with quartz or sapphire windows (Figure 1A). The sample is placed in a sealed sample tube with a compressible cap so that the pressure can be transmitted to the sample with no pressure differential across the tube. Pressure is transmitted to the sample by a hydraulic fluid such as ethanol that will not interfere with the optical detection.

2.1.2. NMR Spectroscopy at High Pressure. Two approaches have been used for variable pressure NMR: in one, only the sample is pressurized; in the other, the NMR probe is pressurized, including the transmitter/detector coils. The former method is simpler in that normal (commercial) NMR probes can be employed; various approaches have been used to pressurize the sample. In one, a quartz tube is sealed at low temperature with a substance that expands and pressurizes the sample when it warms to ambient temperatures (Morishima et al., 1980). In another, the sample is sealed in a small internal diameter quartz tube connected by means of a flexible capillary to a hydraulic pressure generator (Wagner, 1980). Alternatively, the sample can be contained in a folded capillary (Yonker et al., 1995), or in a sapphire tube (Ehrhardt et al., 1996). These studies have been limited to pressures of approximately 1 kbar, although higher pressures should be achievable in principle by these methods.

The second approach requires a custom NMR probe that serves as a pressure bomb (Benedek and Purcell, 1954). Recently, this approach has been used to pressurize NMR samples up to 10 kbar (Ballard et al., 1996). The NMR probe is constructed of a low magnetic susceptibility material with high tensile strength, such as a paramagnetic titanium alloy or a diamagnetic beryllium-copper alloy (Zahl et al., 1994; Frey et al., 1990; Jonas et al., 1993). Such a probe constructed of beryllium-copper has been designed in our laboratory to fit inside a standard probe enclosure for a wide-bore Bruker 400 MHz NMR spectrometer (Mooberry et al, 1996); this design (Figure 1B) supports triple-resonance NMR data acquisition (^{1}H observe, ^{13}C and ^{15}N excitation, with ^{2}H lock) (Mooberry et al., 1997). This design will contain pressures as high as 4.2 kbar. Radio frequency leadthroughs across the high pressure vessel allow for excitation and observation. A hydraulic fluid that contributes minimum NMR signal is used to transmit pressure from the pressure generating system to the sample. Fluids that have proved useful for NMR experiments include carbon disulfide (which has the disadvantages of high vapor pressure, unpleasant odor, and high toxicity) and a fluorocarbon such as Fluorinert (3M Specialty Chemicals Division, St. Paul, MN). We use a custom NMR tube with a movable piston to separate the sample from the pressure transmitting fluid. The piston moves in response to pressure changes to equalize pressure across the glass wall of the sample tube.

2.2. Pressure Generation Systems

The pressure pump and valves are an important part of the overall system for pressure generation, maintenance, and measurement. Our laboratory makes use of two fairly

1A

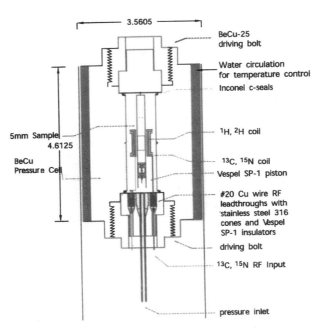

1B

Figure 1. (A) Schematic diagram of high pressure optical cell. Quartz or sapphire windows permit sample excitation and observation. A sample holder with a movable piston or collapsible tube is used to separate the sample liquid from the pressure-transmitting fluid (typically ethanol). (B) High pressure nuclear magnetic resonance probe (Mooberry et al., 1997). A liquid that provides a minimal NMR signal, such as carbon disulfide or a fluorocarbon (see text), is used as a pressure-transmitting fluid so as to reduce background intensity.

standard pressure generation systems: a manual generating pump connected to a series of valves and a Bourdon type pressure gauge (Heise, Model CM: Dresser Industries, New-town, CT) (Figure 2A) and an automated system (Advanced Pressure Products, Ithaca, NY) that uses an electric motor and hydraulic pump, with electric valves and a digital pressure meter (Figure 2B). The latter system is entirely computer controlled and allows for automatic maintenance of pressure even in the presence of slight leaks. The Bourdon and electronic gauges were found to provide equivalent pressure readings over the pressure range between 1 bar and 4.2 kbar.

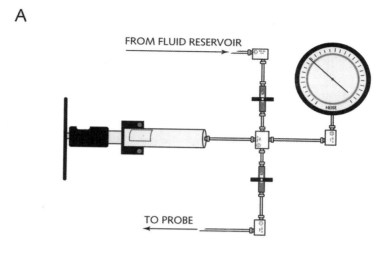

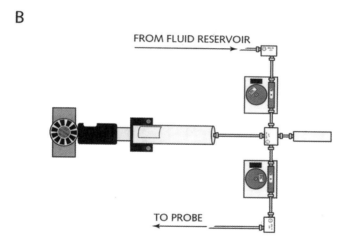

Figure 2. Two systems used in our laboratory to generate pressures up to 4.2 kbar. (**A**) a manual pump and valves with an analog Bourdon-type pressure gauge, and (**B**) an electric motor driven pump with electronic valves and digital pressure gauge. The digital pressure gauge and computer control of the pump allow for excellent mainte-nance of pressure over long time periods.

3. EQUILIBRIUM PRESSURE DENATURATION

In this section we describe equilibrium pressure effects on proteins from the theo-retical and experimental standpoints.

3.1. Theory of Equilibrium Pressure Effects

Equations 1–9 provide the foundation for a description of pressure effects on protein structure. Because the progress of the reaction typically is observed spectrophotometrically (as described in Section 2), it is necessary to relate these observable quantities to the thermodynamic parameters in Eq. 8. Equation 10 serves this function,

$$K_{obs} = \frac{A_i - A}{A - A_f} \tag{10}$$

where A_i and A_f are the intrinsic signal amplitudes of the initial and final states of the protein, respectively, and A is the amplitude at an arbitrary pressure. An implicit assumption in Eq. 10 is that A_i and A_f are independent of pressure; similar assumptions commonly are made for other means of altering protein structure such as temperature or chemical denaturant. Very small pressure dependencies of either of these terms could lead to large errors in the derived thermodynamic parameters.

The relationship between the equilibrium constant, determined from Eq. 10 and the standard free energy change is given by

$$\Delta \overline{G}^{\circ} = -RT \ln K_{eq} \tag{11}$$

Finally, another equation relates the equilibrium constant for the reaction (as determined experimentally from Eq. 10) to the standard volume change,

$$\left(\frac{\partial \ln K_{eq}}{\partial P} \right)_T = -\frac{\Delta V^{\circ}}{RT} \tag{12}$$

In addition, the observable at any pressure, A, is related to the standard free energy change at atmospheric pressure and the standard volume change at atmospheric pressure by

$$A = \frac{A_i + A_f \exp\left(-\dfrac{\Delta \overline{G}^{\circ}_{obs,1bar}}{RT} - \dfrac{p\Delta \overline{V}^{\circ}_{obs}}{RT} \right)}{1 + \exp\left(-\dfrac{\Delta \overline{G}^{\circ}_{obs,1bar}}{RT} - \dfrac{p\Delta \overline{V}^{\circ}_{obs}}{RT} \right)} \tag{13}$$

Typically the standard volume change is assumed to be independent of pressure (i.e. $\ln K_{eq}$ is assumed to be linear as a function of pressure). The consequences of this assumption are discussed below.

To illustrate the effect of pressure on a typical equilibrium between the folded and unfolded states of a protein, Figure 3 shows how the pressure dependence of ΔG° depends on the magnitude of the volume change ΔV° for the reaction. The larger the ΔV°, the greater the rate of change of ΔG° with pressure. This means that smaller pressure ranges are required to probe reactions with large ΔV°, and that larger pressure ranges are needed to observe reactions with small ΔV°. The transition pressure can be tuned to the accessible pressure range by changing the solution conditions to alter $\Delta \overline{G}^{\circ}_{obs,1bar}$. The protein stability can be decreased by changing the pH or temperature (Zipp & Kauzmann, 1973), by adding denaturants such as urea or guanidinium chloride (Vidugiris et al., 1996), or by adding stabilizers such as xylose (Frye et al., 1996; Frye & Royer, 1997).

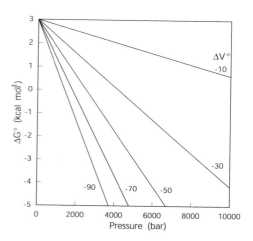

Figure 3. The pressure dependence of the standard volume change of a reaction $\Delta V°$ determines the pressure effect on the position of the equilibrium. This figure, which neglects compressibility changes, shows how the pressure dependence of $\Delta G°$ for a reaction (for which $\Delta G°_{1\,bar} = 3$ kcal mol^{-1} at 25 °C) depends on the magnitude of $\Delta V°$ for the reaction (values indicated in the figure). Typical $\Delta V°$ values for protein denaturation are on the order of -15 to -90 mL mol^{-1}; typical values for subunit dissociation are much larger (Table II).

3.2. Equilibrium Pressure Denaturation of Ribonuclease A Using High Pressure NMR

Bovine pancreatic ribonuclease A (RNase A) is one of the best characterized proteins that exhibits reversible pressure denaturation. The protein is relatively small, monomeric, and well behaved (it is soluble and does not aggregate under most conditions). In addition, it has a highly resolved histidine $^1H^{\varepsilon 1}$ NMR spectrum whose peaks have been assigned (Markley, 1975) as shown in Figure 4.

Because the lifetimes (in seconds) of the interconverting native and denatured states of RNase A are long compared to the inverse of the difference between the chemical shifts of the histidine peaks for these two forms (in Hz^{-1}), each form of the protein gives rise to a distinct sets of signals. In native RNase A, each of the four histidines is located in a different chemical environment and thus has a characteristic chemical shift. In denatured RNase A, however, the side chains are all solvated, and their chemical environments are quite similar. The populations of the folded and unfolded protein can be determined by evaluating the intensities of the respective peaks from these species relative to the intensity of the peak from an internal reference compound. Equation 10 thus can be used to determine the equilibrium constant for denaturation at each pressure investigated. The well resolved resonances from the $^1H^{\varepsilon 1}$ of each of the four histidine residues serve as convenient signals for this analysis. A denaturation profile can be constructed from the pressure dependence of the area of one of the signals, as illustrated in Figure 5 by an analysis of the $^1H^{\varepsilon 2}$ signal from histidine 12 in native RNase A.

3.2.1. Determination of the Volume Change from Pressure Denaturation Data. To measure the standard free energy and volume changes at atmospheric pressure, the experimental data are fitted to Eq. 12, which describes the pressure dependence of a spectroscopic observable. The analysis of many types of spectroscopic data by Eq. 10 requires a four-parameter fit, with $\Delta \overline{G}°_{obs,1bar}$, $\Delta \overline{V}°_{obs}$, A_i, and A_f as the adjustable parameters. A distinct advantage of NMR data is that A_i, and A_f are known independently. For example in protein denaturation, signals from the folded form have $A_i = I$ and $A_f = 0$, and signals from the denatured form have $A_i = 0$ and $A_f = I$, where I is the unit NMR signal intensity for the group being observed. Therefore, for NMR data, the only the adjustable parameters in this

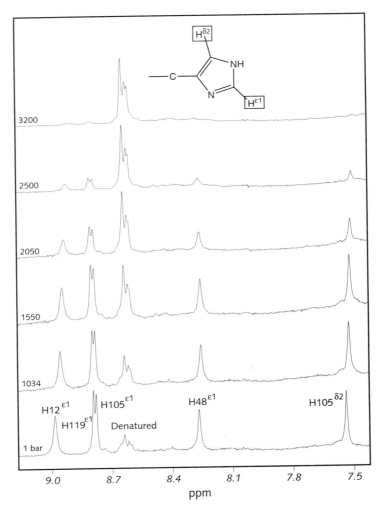

Figure 4. 400 MHz ^{1}H NMR spectra of bovine pancreatic ribonuclease A (RNase A) at pH 2.0 and 295 K. Separate signals are observed for a nucleus in the folded and unfolded states of the protein (condition of slow exchange on the NMR chemical shift time scale). (Data from K. E. Prehoda, E. S. Mooberry, & J. L. Markley, unpublished).

model are the standard free energy $\Delta \overline{G}^{\circ}_{obs,1bar}$ and the volume change at atmospheric pressure $\Delta \overline{V}^{\circ}_{obs}$. The results of nonlinear least squares analysis of the experimental NMR data according to Eq. 13 are shown by the curve in Figure 5. The standard volume change for the denaturation of RNase A, like that of other proteins, is remarkably small. The physical basis for standard volume changes accompanying protein unfolding is discussed below in Section 5.

3.2.2. Compressibility Change for Protein Denaturation. Although it is clear that both the folded and unfolded forms of a protein are compressible (i.e., their partial molar volumes are pressure dependent), it generally is assumed that these compressibilities are

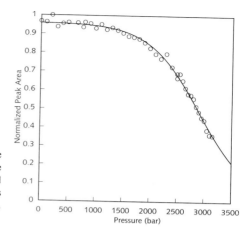

Figure 5. Pressure dependence of the intensity of the NMR signal from $^1H^{\epsilon l}$ of histidine 12 in the folded state (circles). The curve represents a fit of the experimental data to the model embodied in Eq. 13. The results of this analysis are shown in Table III. (From K. E. Prehoda, E. S. Mooberry, & J. L. Markley, unpublished).

equal and thus do not affect the volume change for the reaction $\Delta V°$. The pressure dependence of the volume change is known as the standard compressibility change $\Delta K°$

$$\Delta K° = (\Delta V°/\, p)_T \qquad (14)$$

Little attention has been paid to evaluating the validity of the assumption that the compressibility change for protein denaturation is zero or to the possible effects the neglect of a non-zero compressibility change would have on the accuracy of other thermodynamic parameters that characterize the process. We have attacked this problem by considering two questions: 1) If test data generated from a theoretical model that includes a compressibility change are evaluated by conventional curve fitting to Eq. 13, which neglects a compressibility change, will it be apparent that an oversimplified model is being used? 2) Can experimental data be evaluated in a way that permits the determination of the compressibility change for protein denaturation?

To test the first question, synthetic data were constructed according to a model that included nonzero compressibility changes. As shown in Figure 6, analysis of these synthetic data by a four-parameter fit with a compressibility change of zero yielded apparent agreement. The disagreement became noticeable, however, when a two-parameter fit (in which the baseline values were fixed) was used. In this, case, the fit could be improved by including the compressibility change as a third adjustable parameter. The results of this model analysis suggests that it will be difficult to determine compressibility changes from spectroscopic data such as fluorescence that require fitting of the spectral plateaus but that NMR spectroscopy may offer the means for such analysis. Significantly, the standard free energy and volume changes returned from such an analysis can be affected dramatically by the assumption of zero compressibility change. The magnitudes of these errors can vary widely, depending on the actual compressibility change and the conditions of the experiment. For example, in Figure 6A, the assumption that $\Delta K° = 0$ results in systematic errors of 1 kcal mol^{-1} in $\Delta G°$ and 42 mL mol^{-1} in $\Delta V°$.

To address the second question we must examine actual experimental data such as those shown in Figure 5. With NMR data, the intrinsic signal amplitudes need not be included as adjustable parameters in the analysis. This allows us to include $\Delta K°$ in the fit of the experimental data such that it can be determined reliably. The analysis of the pressure denaturation data for RNase A with and without compressibility change is shown in Table

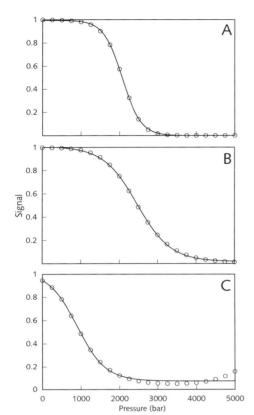

Figure 6. Synthetic spectroscopic data sets for pressure denaturation generated with different values of compressibility (circles) are shown along with curves (—) that represent the best fits to these data by the model (Eq. 10) that ignores compressibility and requires fitting of the signal intensity at high and low pressure. (**A**) *Synthetic data:* $\Delta G°$ = 4 kcal mol^{-1}, $\Delta V°$ = −60 ml mol^{-1}, $\Delta K°$ = 0.02 ml mol^{-1} bar^{-1}. *Model:* $\Delta G°$ = 5 kcal mol^{-1}, $\Delta V°$ = −102 ml mol^{-1}. (**B**) *Synthetic data:* $\Delta G°$ = 4 kcal mol^{-1}, $\Delta V°$ = −80 ml mol^{-1}, $\Delta K°$ = −0.01 ml mol^{-1} bar^{-1}. *Model:* $\Delta G°$ = 3.3 kcal mol^{-1}, $\Delta V°$ = −56 ml mol^{-1}. (**C**) *Synthetic data:* $\Delta G°$ = 1.7 kcal mol^{-1}, $\Delta V°$ = −85 ml mol^{-1}, $\Delta K°$ = −0.025 ml mol^{-1} bar^{-1}. *Model:* $\Delta G°$ = 1.4 kcal mol^{-1}, $\Delta V°$ = −66 ml mol^{-1}. All values are for 298 K and 1 bar. (Prehoda, 1997).

Table III. Effect of Compressibility Change on the Pressure Denaturation Parameters for Bovine Pancreatic Ribonuclease A[a]

Histidine ^{1}H$^{\epsilon 1}$ signal(s) used in the analysis[b]	$\Delta G°_{1\,bar}$ (/kcal mol^{-1})	$\Delta V°_{1\,bar}$ (/mL mol^{-1})	$\Delta K°_T$ (/mL mol^{-1} bar^{-1})
$\Delta K°_T$ fixed at zero			
H12	3.7 ± 0.1	−53 ± 1	0
H48	3.9 ± 0.6	−53 ± 5	0
H119+H105	3.7 ± 0.2	−53 ± 2	0
Denatured	3.70 ± 0.08	−56 ± 1	0
$\Delta K°_T$ included in fit			
H12	2._1 ± 1	−6 ± 13	0.01_7 ± 0.01
H48	2._4 ± 2	−5 ± 19	0.01_6 ± 0.01
H119+H105	2._1 ± 2	−6 ± 15	0.01_7 ± 0.01
Denatured	2.33 ± 0.07	−8 ± 2	0.019 ± 0.001
Ln K_{obs} [c]	2.22 ± 0.05	−13 ± 3	0.015 ± 0.002

[a]From K. E. Prehoda. E. S., Mooberry, & J. L. Markley, manuscript in preparation.
[b]Peaks from histidine residues in the folded protein (H12, H48, and H119+H105, which are overlapped); aggregate signals from the unfolded protein (denatured). The larger and sharper signals from the histidines in the unfolded protein provide fits with lower uncertainty than those of the folded protein and thus are expected to provide the most reliable thermodynamic values.
[c]The parameter ln K_{obs} could not be fitted reliably without including the compressibility change as a parameter.

III. The standard deviations for the parameters are somewhat larger when $\Delta K°$ is included because more parameter sets will fit the data with roughly equal χ^2 values.

The reliability of the compressibility change measurement can be improved by determining the standard free energy change $\Delta G°$ by an independent method. As described above, the standard free energy and volume changes can be extremely sensitive to the compressibility change. For example, from differential scanning calorimetry, $\Delta G°$ for unfolding of RNase A at the conditions of the experiment shown in Figure 5 and atmospheric pressure, is 2.4 kcal mol^{-1} (Makhatadze et al., 1995). When the compressibility change is assumed to be zero in the pressure denaturation experiment, a standard free energy change of 3.7 kcal mol^{-1} models the data well. Only when a compressibility change of 0.015 mL mol^{-1} bar^{-1} is included in the analysis does the standard free energy change agree with the calorimetry results (2.2 kcal mol^{-1}). Care must be taken, however, to insure that both methods are following the same reaction. For example, certain proteins aggregate at high temperature which could result in a standard free energy change derived from calorimetry different from that observed in a pressure denaturation experiment.

4. KINETICS OF PRESSURE DENATURATION: PRESSURE RELAXATION STUDIES

Thus far only the equilibrium aspects of the denaturation process have been discussed. Investigation of pressure effects on rates can provide additional insight into the mechanisms of protein structural changes including folding and unfolding. How a protein adopts its specific native conformation from the random coil has recently taken on medical importance as certain diseases, for example, prion disease (Harrison, et al., 1997), appear to be caused by incorrect folding of proteins. Among the powerful techniques have emerged in the past few decades for studying the mechanism of protein folding, NMR has the distinct advantage of being capable of providing information at atomic resolution throughout the protein structure. For example, equilibrium hydrogen exchange and pulse labeling (Bai et al., 1995) can often be followed for every backbone amide proton in a protein using ^{15}N-^{1}H correlation spectroscopy.

Whereas the volume change for the reaction $\Delta V°$ is the parameter that emerges from an analysis of the pressure dependence of equilibria, the parameter derived from the pressure dependence of rates is the activation volume for the reaction $\Delta V^{\ddagger}$. A common approach is pressure relaxation, in which the pressure is changed rapidly and the progress to equilibrium is followed as a function of time. This technique has been used successfully to study the kinetics of ethidium bromide binding to DNA (Macgregor et al., 1985), the conversion between metarhodopsin I and II (Attwood & Gutfreund, 1980), and the assembly of myosin filaments (Coates et al., 1985; Davis, 1981; Davis, 1985). In these studies, the pressure was changed rapidly by means of a piezoelectric crystal, and the relaxation to equilibrium was followed over the time scale of several seconds. From the pressure dependence of the rate constants, it was possible to estimate the volume of the activated state for folding relative to the native and denatured states. These activation volumes can then be used to evaluate models for the transition state for the folding reaction.

4.1. Theory of Pressure Relaxation Kinetics

What can we learn from following the relaxation to equilibrium after a rapid pressure jump? This question can be answered by examining the pressure dependence of the rate constants. The folding ⇌ unfolding reaction for many proteins can be modeled adequately by a two-state equilibrium, because no populated intermediates accumulate. Thus,

$$N \underset{k_{-1}}{\overset{k_1}{\rightleftharpoons}} D \tag{15}$$

For the two-state equilibrium given in Equation 15, the time dependence of the native and denatured protein concentrations is given by

$$\frac{d[D]}{dt} = -\frac{d[N]}{dt} = k_1[N] - k_{-1}[D] \tag{16}$$

To examine the effect of a pressure perturbation on this reaction, it is useful to define a progress variable Δ, which is the deviation of reactants and products from their equilibrium concentrations

$$\Delta = [N] - [N]_{eq} = [D]_{eq} - [D] \tag{17}$$

The time dependence of the progress variable as determined from the rate law (Eq. 15), is then

$$\frac{d\Delta}{dt} = \frac{d[N]}{dt} = -k_1([N]_{eq} + \Delta) + k_{-1}([D]_{eq} - \Delta) \tag{18}$$

At equilibrium, the concentrations of N, $[N]_{eq}$, and D, $[D]_{eq}$, are time independent and satisfy the equation,

$$-k_1[N]_{eq} + k_{-1}[D]_{eq} = 0 \tag{19}$$

Using Eq. 19, we can integrate Eq. 18 between time zero and time t,

$$\int_{t=0}^{t} d\Delta = \int_{t=0}^{t} -k_1([N]_{eq} + \Delta) + k_{-1}([D]_{eq} - \Delta)dt \tag{20}$$

The progress variable is then expressed by its initial value Δ_0 and the sum of the rate constants at time t.

$$\ln\frac{\Delta}{\Delta_0} = -(k_1 + k_{-1})t \tag{21}$$

Equation 21 allows us to determine the folding and unfolding rate constants spectroscopically because the progress variable can be related to an observable that is responsive to the concentration of native or denatured species. For example,

$$\frac{\Delta}{\Delta_0} = \frac{A_N - A_{N,eq}}{A_{N,0} - A_{N,eq}} \tag{22}$$

where A_N is the amplitude of the native species at time t, A_N at equilibrium, and $A_{N,0}$ at time zero. A similar expression can be derived in terms of A_D. Thus, the signal amplitude as a function of time can be modeled to evaluate the sum of the forward and reverse rate constants,

$$A_N = A_{N,eq} + (A_{N,0} - A_{N,eq})e^{-(k_1 + k_{-1})t} \tag{23}$$

Additional information is needed in order to extract the individual rate constants k_1 and k_{-1}, from Eq. 23. The equilibrium constant, $K_{obs} = k_1/k_{-1}$, can be used to deconvolute the individual rate constants. Alternatively, if the magnitudes of k_1 and k_{-1} are sufficiently different, they can be evaluated graphically by plotting the logarithm of the relaxation time τ ($1/(k_{-1} + k_1)$) against pressure; the slopes of the linear portions of the curve before and after the denaturation midpoint yield k_{-1} and k_1, respectively, assuming that $k_1 \ll k_{-1}$ below the transition and $k_1 \gg k_{-1}$ at pressures above the transition.

Finally, the pressure dependence of the rate constants allows us to determine the activation volumes for folding and unfolding through transition state theory,

$$\left(\frac{\partial \ln k_1}{\partial P}\right)_T \cong -\left(\frac{\partial \Delta G_1^{\circ\ddagger}/RT}{\partial P}\right)_T = -\frac{\Delta V_1^{\circ\ddagger}}{RT} \tag{24}$$

4.2. Pressure Relaxation Studies of Staphylococcal Nuclease

Staphylococcal nuclease is a 16 kD single domain, monomeric protein that has been used extensively as a model for protein folding. Wild type and mutant forms of the enzyme have been studied with high pressure fluorescence and NMR. At 1750 bar and 25°C the native and denatured forms of wild type nuclease have comparable populations.

Pressure-jump fluorescence spectroscopy (Figure 7) has been used to determine the activation volumes for folding ($+90$ mL mol^{-1}) and unfolding ($+20$ mL mol^{-1}) of wild type staphylococcal nuclease (Vidugiris et al., 1996). Figure 8 shows a schematic representation of the reaction coordinate with the volume as the vertical axis and with cartoons representing the folded, transition, and unfolded states. This approach also has been used to investigate effects of proline to glycine mutations on folding and unfolding rates (Vidugiris et al., 1996).

The kinetics following a pressure jump also can be observed also by NMR spectroscopy. Interestingly, because of the large activation volumes for folding and unfolding, at pressures near the midpoint of the transition the kinetics of unfolding following a positive pressure jump and the kinetics of unfolding following a negative pressure jump are sufficiently slowed that they can be monitored by NMR in real time. Instead of being in the millisecond range as they are at atmospheric pressure, the half times are on the order of several minutes near the midpoint pressure for the reaction. Figure 9 shows the intensity of the histidine $^1H^{\epsilon 1}$ signals from denatured staphylococcal nuclease as a function of time following positive and negative pressure jumps.

5. DISCUSSION OF VOLUME CHANGES THAT ACCOMPANY PROTEIN DENATURATION

The impetus for using pressure to study proteins is to determine the volumetric characteristics of the denaturation reaction. However, the origin of the volume change is con-

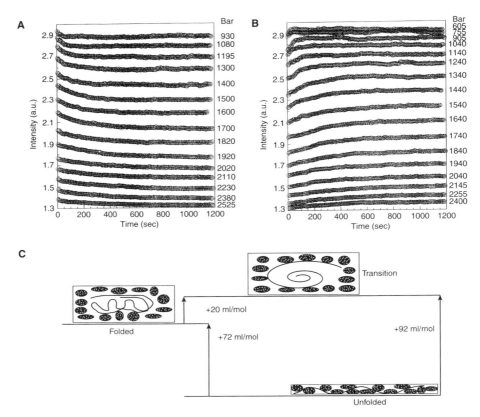

Figure 7. Pressure-jump fluorescence spectroscopic investigation of protein folding of staphylococcal nuclease. The intensity of the intrinsic tryptophan emission of nuclease as a function of time following (**A**) positive (unfolding) and (**B**) negative (folding) pressure jumps are shown as relaxation profiles. The fluorescence intensity (in arbitrary units) is shown at the left, and the equilibrium pressure reached at the end of the jump is shown at the right of each panel. Solid lines are fits to the data obtained by time-domain fluorescence global analysis. (**C**) The natural logarithm of the relaxation time ln τ, derived from this analysis at each pressure is plotted as a function of pressure. The slope of the linear portion of the low-pressure limb of the curve is a measure of the activation volume for protein unfolding, and the slope of the linear portion of the high-pressure limb of the curve is a measure of the activation volume for protein folding (Figure adapted with permission from Vidugiris et al., 1995).

troversial. In this section we examine the detailed components of this thermodynamic parameter and summarize our approach to their analysis.

5.1. Equilibrium Volume Change for Protein Denaturation

The most surprising result from pressure denaturation experiments is the remarkably small difference in the partial molar volumes of the folded and unfolded states. For example, from the experiments discussed in the previous section, the volume change for unfolding of RNase A at atmospheric pressure, pH* 2.0 and 21 °C is only about -6 mL mol^{-1}. This volume change constitutes only a small fraction (0.07%) of the partial molar volume of native RNase A under these conditions (8800 mL mol^{-1}) (Tamura & Gekko, 1995). Additionally, it appears that ΔV°_{obs} is roughly independent of molecular weight: large pro-

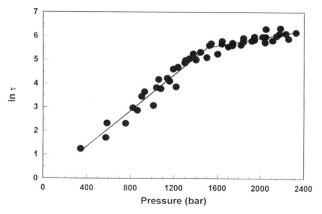

Figure 8. Schematic model consistent with the experimental volume changes. Small ovals represent water molecules; a line represents the polypeptide chain. The partial molar volume (protein plus associated solvent) is smallest for the unfolded protein, in part, owing to minimization of packing defects. The partial molar volume of the folded protein is of intermediate magnitude, and that of the transition state for folding/unfolding is the largest. This implies that the protein structure is sufficiently swollen in the transition state to lead to increased packing defects, but not enough to permit solvent to penetrate into the structure to occupy these cavities (Figure reproduced with permission from Vidugiris et al., 1995).

teins, although having much larger partial molar volumes than small proteins, show similar volume changes for unfolding (Heremans, 1982).

It is instructive to examine the detailed physical phenomena that give rise to the observed volume changes for pressure-induced reactions involving proteins. These include conformational changes in the native state, subunit dissociation, ligand dissociation, and protein denaturation. As a simplification, the following will concentrate on volume changes that attend protein denaturation; it should be borne in mind that the principles apply to a wide range of reactions. An earlier version of the analysis presented here has been published (Prehoda & Markley, 1996).

Our approach has been to start by analyzing volume effects in model systems. When considering volume effects in the denaturation reaction, the volume of the protein itself, separate from the solvent system, undoubtedly is expected to differ in the folded and unfolded states. The high degree of compactness of folded proteins is often cited as one of their defining features (Dill, 1990; Richards, 1977). Although highly compact, the folded

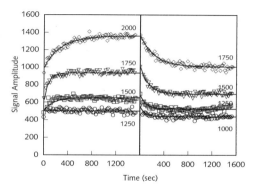

Figure 9. Pressure relaxation of staphylococcal nuclease monitored by NMR at pH 5.3, 298 K. The denatured histidine signal from staphylococcal nuclease is shown as a function of time for several small pressure perturbations. The solid lines represent best fits of the data to single exponentials (K. E. Prehoda, E. S. Mooberry, R. A. Chylla, & J. L. Markley, unpublished).

structure contains many voids and packing imperfections that are inaccessible to solvent (Richards, 1977). By contrast, the unfolded protein exhibits little packing of the peptide chain, and the packing imperfections that were present in the folded state are expected to become occupied by solvent (Weber & Drickamer, 1983). The volume occupied by the protein molecule itself (neglecting its solvation) in the unfolded state is expected to be considerably smaller than that in the folded state, owing to such packing differences.

Solvation of uncharged groups is another important component of the partial molar volumes of the folded and unfolded states. Since protein unfolding is accompanied by a large increase in solvent-exposed, nonpolar surface area, the transfer of liquid hydrocarbons into water has been invoked as a model for protein denaturation. The pressure dependence of the solubility of liquid hydrocarbons in water provides information in the form of a transfer volume ΔV°_{tr}. Results from a number of hydrocarbons indicate that ΔV°_{tr} is approximately -2.0 ml mol^{-1} methylene^{-1} at atmospheric pressure and 25°C (Sawamura et al., 1989). Thus, the transfer of a pure liquid hydrocarbon into water under these conditions is accompanied by a decrease in volume. This implies that hydration of apolar solutes is more dense than bulk water and that the exposure of apolar surfaces leads to a decrease in volume (Dill, 1990; Heremans, 1982; Weber & Drickamer, 1983). On extrapolation of this model to protein unfolding, it is expected that hydration of nonpolar surfaces should make a significant negative contribution to ΔV°_{obs}. Indeed, the traditional interpretation of the negative ΔV°_{tr} describes it as a collapse of hydration water around the hydrophobic solute (Dill, 1990).

A third effect on volume is the solvation of charged groups, or electrostriction. It is well known that the solvation of charged groups, particularly positively charged groups, contributes a negative term to ΔV°_{obs}. In protein denaturation, this effect comes to play whenever charged groups that are buried in the native state become exposed to solvent in the denatured form. The magnitude of this effect is likely to be much smaller than the previously discussed contributions because proteins contain few buried charged groups. The solvation of polar groups is expected to be volumetrically similar to water itself such that their exposure should have little effect on the observed volume change (Royer, 1995; Weber & Drickamer, 1983).

Whereas it is very difficult, to determine absolute partial molar enthalpies, entropies, or free energies of solutes from experiments, absolute partial molar volumes can be determined. Because absolute quantities are not dependent on the characteristics of the reference state, it is possible to circumvent problems associated with transfer quantities by using the absolute partial molar volumes of model compounds to rationalize the protein unfolding behavior under high pressure.

5.1.1. Hard Sphere Model of the Partial Molar Volume. To understand the volumetric aspects of proteins, a hard-sphere model can be used to evaluate the contributions to ΔV°_{obs}. This type of model is useful for dissecting the partial molar volume into components arising from molecular volume and solvation effects (Prehoda & Markley, 1996). In this model, the total volume of a solution V_T is divided into components from the solute, solvent, and remaining void volume,

$$V_T = n_1 \overline{V}_1^* + n_2 \overline{V}_2^* + V_{int} \tag{25}$$

where n_1 and n_2 are the number of moles of solvent and solute, $\overline{V}_1^*$ and $\overline{V}_2^*$, respectively, are the molar molecular volumes of solvent and solute, and V_{int} is the volume interstitial to

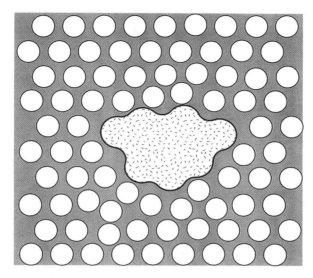

Figure 10. Schematic representation of the hard sphere model of the partial molar volume. The total volume of a solution is represented by the molecular volume of the solvent (circles), the molecular volume of the solute (irregular object), and the voids interstitial to the solvent and solute (shaded) (Adapted with permission from Prehoda & Markley, 1996).

both (Figure 10). The molecular volume is defined as the volume enclosed by the molecular surface, which is the sum of the contact and reentrant areas traced out by a spherical probe rolled over the molecule (Richards, 1977). The difference between the molecular surface and the more commonly used solvent accessible surface is shown in Figure 11. The partial molar volume of the solute in terms of the hard-sphere model is given by

$$\overline{V}_2 \equiv \left(\frac{\partial V_T}{\partial n_2}\right)_{n_1,T,P} = \left(\frac{\partial V_{\text{int}}}{\partial n_2}\right) + \overline{V}_2^*$$

(26)

Equation 26 assumes that the molecular volume of solvent is not perturbed by addition of solute. This simple equation provides an intuitive picture for the different compo-

Figure 11. Molecular and solvent accessible surfaces of a solute. The molecular surface is defined as the sum of the contact surface (those parts of the molecular van der Waals surface that can actually be in contact with the surface of the probe) and the reentrant surface (the interior-facing part of the probe when it is simultaneously in contact with more than one atom). The solvent accessible surface is the continuous sheet defined by the locus of the center of the probe (Richards, 1977). Both surfaces are highly dependent on the radii of the probe and the atoms of the molecule for which the surface is being calculated. A probe radius of 1.4 Å is commonly used to approximate a water molecule.

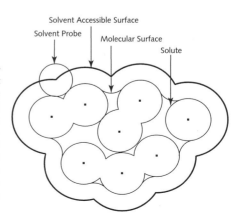

nents of the partial molar volume. When a solute is added to a solution, the total volume will be altered as the result of two effects: the volume that is displaced by the solute itself ($\overline{V}_2^*$) and the volume from interactions between the solute and solvent ($\overline{V}_{int}$), which can be positive or negative.

Recall from Eq. 8 that the volume change for unfolding is the difference in partial molar volumes of the unfolded and folded states. Using Equation 26 and dissecting the partial molar interstitial volume component into contributions from polar and nonpolar surfaces, the volume change for protein unfolding can be written

$$\Delta V^{\circ}{}_{obs} \approx \Delta \overline{V}_2^* + \Delta \overline{V}_{int,np} + \Delta \overline{V}_{int,p} \tag{27}$$

where $\Delta \overline{V}_2^*$ is the difference in solvent-excluded molecular volumes between the unfolded and folded states, and $\Delta \overline{V}_{int,np}$ and $\Delta \overline{V}_{int,p}$ are the contributions from hydration of nonpolar and polar surfaces, respectively. It is assumed that each component of the hydration, polar and nonpolar, are additive, such that there is no cooperativity between polar and nonpolar hydration. Undoubtedly this assumption is false, but it has been used successfully in the approximation of contributions of polar and nonpolar hydration to other thermodynamic quantities, such as the heat capacity change for unfolding (Makhatadze & Privalov, 1993; Spolar et al., 1992). Additionally, electrostriction of water molecules by charged groups on the protein is neglected. Electrostriction, the binding of water dipoles by charged species, is accompanied by a volume decrease (Morild, 1981). Although proteins can be highly charged molecules, they contain relatively few buried charges. Therefore, as noted above, the amount of electrostricted water should be roughly the same in both the folded and unfolded states. Although the contribution of electrostriction to the partial molar volumes of each individual state may be significant, the net contribution to $\Delta V^{\circ}{}_{obs}$ is assumed here to be small enough to be neglected as a first approximation.

5.1.2. Partial Molar Volumes of Model Compounds. The components of the partial molar volumes of model compounds were determined as follows (Prehoda & Markley, 1996): overall volumes were taken from experimental partial molar volumes in water, all determined at atmospheric pressure and 25 °C (Masterton, 1954; Shahidi, 1981; Shahidi et al., 1977). Molecular volumes were calculated from x-ray structures from the Protein Data Bank (http://www.pdb.bnl.gov) using the program PQMS (Connolly, 1985; Connolly, 1993). The solvent-excluded molecular volumes were calculated with a probe radius of 1.4 Å. The partial molar interstitial volumes of nonpolar compounds were evaluated by subtracting the calculated molecular volume from the experimental partial molar volume for each solute. In order to determine the amount of interstitial volume per unit of surface area, $\overline{V}_{int}$ was normalized to the solvent accessible surface area of the solute.

The solvation of model nonpolar solutes, according to the hard-sphere model described above, can be approximated using experimental partial molar volumes and calculated molecular volumes. These two components are shown in Figure 12A for a series of hydrocarbons as a function of their solvent accessible surface area. It is evident that the partial molar volumes of these hydrocarbons are significantly larger than their molecular volumes. The difference between these two quantities is the partial interstitial volume $\overline{V}_{int}$ which is plotted for this series of hydrocarbons in Figure 12B as a function of solvent accessible surface area. The slope of this line, 0.166 ml mol^{-1} Å^{-2}, is the amount of void introduced into the solution per square angstrom of nonpolar surface area.

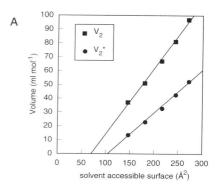

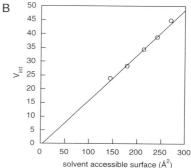

Figure 12. Analysis of partial molar volumes of a series of hydrocarbons according to the hard sphere model (Prehoda & Markley, 1996). (**A**) Experimental partial molar volumes and molecular volumes for the series of hydrocarbons in water. The experimentally determined partial molar volumes (Masterton, 1954), of methane, ethane, propane, benzene and toluene are shown (squares) as a function of solvent accessible surface. The excluded volumes of the same solutes, as calculated by the program PQMS (Connolly, 1993), are also shown (circles). (**B**) Partial molar interstitial volume for the same series of hydrocarbons. The difference in the experimental partial molar volumes and the calculated molecular volumes, termed the partial molar interstitial volume, is shown for the same hydrocarbons as in A. The slope of the line through the points represents the amount of void introduced into the solution per square angstrom of surface area.

The partial molar interstitial volume for the polar compounds was calculated in a similar fashion, except that the contribution from the nonpolar portions of the amides was subtracted using the normalized values from the hydrocarbons. However, data for these compounds are more limited, and the values contain greater error resulting from the need to subtract the calculated contribution from the nonpolar component. Thus, the partial molar interstitial volume for hydration of polar groups is less well determined than that for nonpolar groups. The results of this analysis are shown in Table III. In contrast to solvation of nonpolar surface area, which introduces a significant amount of void into the solvent, the solvation of polar surface introduces relatively little void.

5.1.3. Components of the Volume Change for Protein Denaturation. The molecular volumes for several proteins were calculated in a manner identical to that used with the model compounds (Prehoda & Markley, 1996). Calculations of the molecular volumes of the native state made use of x-ray structures (at highest available resolution) from the Protein Data Bank. To approximate the molecular volume of the denatured state, an extended chain was used. This approximates the lower volume limit for a random, expanded chain; the actual volume would be slightly higher, because local structure would prevent penetration of the solvent. $\Delta \overline{V}_2^*$ was then determined from the difference of the solvent-excluded molecular volume of the extended chain and that of the native state structure.

The nonpolar and polar surface areas of these molecular models were calculated by use of the PQMS program. These were then multiplied by the normalized partial molar interstitial volumes of the nonpolar and polar model compounds, respectively, to yield the hydration components of the volume change, $\Delta \overline{V}_{int,np}$ and $\Delta \overline{V}_{int,p}$. For example,

Table IV. Calculated and Experimental Volume Parameters for Proteins[a]

Protein	M_r ($/10^3$)	$\Delta \bar{V}_2^\circ$ ($/10^3$ mL mol^{-1})	$\Delta \bar{V}_{\text{int},np}$ ($/10^3$ mL mol^{-1})	$\Delta \bar{V}_{\text{int},p}$ ($/10^3$ mL mol^{-1})	$\Delta V^\circ_{\text{calc}}$ (/mL mol^{-1})	$\Delta V^\circ_{\text{obs}}$ (/mL mol^{-1})
Ribonuclease A	13.7	−1.39	0.948	0.217	−225	−35
Lysozyme	14.3	−1.79	1.13	0.423	−237	−20
Staphylococcal nuclease	16.8	−1.85	1.28	0.284	−286	−60
Chymotrypsinogen	25.7	−3.52	2.37	0.705	−445	−30

[a]From Prehoda & Markley (1996).

$$\Delta \bar{V}_{\text{int},np} = \frac{\bar{V}_{\text{int},np}}{A_{np}} \Delta A_{np,protein}$$

(28)

The components of the volume change for denaturation, in terms of the model described above, are shown in Table IV for four proteins that have been studied with pressure and for which high resolution structures exist. The triangulation of the surface of
staphylococcal nuclease, as used by the program PQMS to calculate molecular volumes, is
shown in Figure 13. The difference between the molecular volume of the unfolded state, as
approximated by an extended chain, and the folded state is a large negative volume. The
contribution of hydration to the volume change, as calculated from the difference in surface area and the model compound parameters, is a large positive volume. The overall volume change ΔV° is the difference between these two quantities. Although the values
obtained from the model are only approximations, they mimic features of experimental
ΔV° values for protein denaturation in that they are small and negative and show no systematic dependence on the molecular weight of the protein.

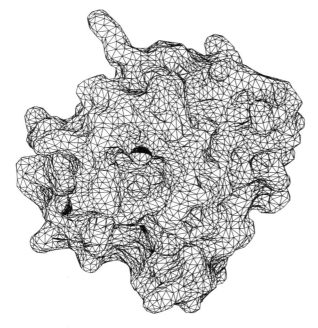

Figure 13. Triangulation of the molecular
surface of staphylococcal nuclease. The solvent-excluded molecular volumes for the
native proteins were calculated from atomic
coordinates derived from high resolution x-
ray crystallography using the PQMS computer program (Connolly, 1985).

5.2. Reevaluation of the Volume Effects of Protein Hydration

Analysis of the partial molar interstitial volume for the series of aqueous hydrocarbons studied here, indicates that the void volume is introduced from solvation of nonpolar surface is approximately 0.166 ml mol^{-1} Å^{-2}. The positive sign and magnitude of this result indicates that the microscopic density of the water of solvation for these solutes is significantly *lower* than that of bulk solvent. This conclusion is contrary to that derived from the negative ΔV°_{tr} values for hydrocarbons, namely that hydrophobic hydration is *more* dense than bulk water. The discrepancy arises because the transfer volume is a function, not only of the partial molar volume of the aqueous hydrocarbon, but also of the molar volume of the reference state (i.e., the pure liquid hydrocarbon). Owing to the relatively low density of liquid hydrocarbons, it is possible that the effect of hydration is masked by the void that is removed from the pure liquid state when the solute is transferred from the pure liquid to the aqueous state. This hypothesis is supported by the pressure dependence of ΔV°_{tr} (Figure 14). Whereas ΔV°_{tr} is negative at atmospheric pressure, it increases with increasing pressure and eventually becomes positive at approximately 2 kbar (Sawamura et al., 1989). At higher pressures, where the low density hydrocarbon is more compressed, the effect of the solute-solvent interaction in the aqueous state dominates, and ΔV°_{tr} becomes positive.

The above analysis suggests that observed volume changes for protein denaturation result from compensation between molecular volume effects and differential hydration of the folded and unfolded states. Whereas the change in molecular volume is negative, as expected, the volume change resulting from increased hydration of the polymer chain in the unfolded state is positive, opposite from that predicted from previous interpretations of transfer volume experimnets. Volume changes calculated according to the present analysis show qualitative agreement with experiment, although the many assumptions made do not support qualitative comparisons.

The striking similarity of the volume changes for unfolding of proteins of similar size may be a result of similarities in the ratio of nonpolar to polar surface area and compactness. Experimental pressure denaturation results indicate that ΔV°_{obs} also is roughly independent of protein size (Table I). The proposed compensation between molecular volume and hydration effects is consistent with this observation. Larger proteins, while having a significantly greater molecular volume difference in the folded and unfolded state, also expose more surface area when unfolding. Thus both terms, while increasing in magnitude with increasing

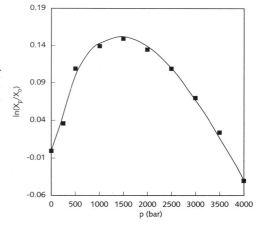

Figure 14. Pressure dependence of the solubility of benzene in water. The natural logarithm of the solubility of benzene (expressed as the mole fraction of benzene at pressure p divided by the mole fraction at zero pressure) is shown plotted as a function of pressure from data in the literature (Sawamura et al., 1989). The transfer volume change ΔV°_{tr} is negative at atmospheric pressure, as indicated by the positive slope. However, as the pressure is increased, the pure liquid hydrocarbon compresses such that ΔV°_{tr} becomes less negative and eventually turns positive at approximately 2 kbar.

molecular weight, compensate one another as the result of their opposite sign, such that the observed volume change has a small molecular weight dependence.

5.3. The Physical Basis of the Compressibility Change on Protein Denaturation

Perhaps the greatest task for the biophysical chemist lies in interpreting thermodynamic quantities from protein denaturation experiments. Although challenging, this task can be highly rewarding, particularly if it reveals elusive principles about protein structure and function. In this section we discuss the physical origins for the positive compressibility change observed for protein denaturation (K. E. Prehoda, E. S. Mooberry & J. L. Markley, manuscript in preparation). The compressibility change determined for RNase A (Table III) indicates that the standard volume change is very small at atmospheric pressure, but becomes more negative at higher pressures. What physical information does this provide about the protein? To begin with, recall from Eq. 9 that we are dealing with partial molar quantities. As a consequence, not only will the solute (protein) contribute to the thermodynamic parameter, in this case the compressibility change, but also the interaction between the solute and solvent (water). Our goal is to deconvolute these factors in order to assess their relative importance. A starting point for understanding the compressibility change on protein denaturation is to decompose this into a term from the protein molecule itself and one from the solvent that interacts with the protein (the protein hydration layer). Implicit in this analysis is the assumption that the compressibility of bulk water (water molecules that do not interact with the protein) is the same when the protein is folded or unfolded. Thus,

$$\Delta K^\circ = n_1 \overline{K}_1^\circ + n_{int} \overline{K}_{int}^\circ \tag{29}$$

A starting point for understanding the compressibility of a folded protein, information about $\overline{K}_1^\circ$ comes from the results of protein crystallography at variable pressure. Comparison of x-ray structures of hen egg white lysozyme determined at 1 kbar and 1 bar showed that the protein itself, separate from the solvent system, is 44 mL mol^{-1} smaller at the higher pressure (Kundrot & Richards, 1987). From this value, we can estimate the compressibility of the folded protein to be 0.04 ± 0.01 mL mol^{-1} bar^{-1}. This effect can be accounted for by compression of the many small, solvent-inaccessible voids that result from packing imperfections in the native state. These are not present in the denatured form, and because covalent bonds are highly incompressible, the compressibility of a fully unfolded protein is assumed to be negligible. This suggests that, the compressibility of the protein itself is significantly higher in the folded form and that the compressibility change in the protein itself upon denaturation is negative ($\overline{K}_1^\circ < 0$). Support for this argument comes from the finding that the chemical shifts of some NMR signals from folded proteins show a pressure dependence consistent with compression (upfield shifts for hydrophobic groups; downfield shifts for hydrogen bonded groups) whereas signals from the unfolded protein do not (Prehoda, 1997; Akasaka et al., 1997). Further investigation of the compressibility of the unfolded protein molecule is warranted, however, particularly because of widespread evidence for residual structure in proteins unfolded by a variety of methods.

What about compressibility effects from the solute-solvent interaction? Here we are concerned with how interactions with the protein affect the normal compressibility of bulk water. The solute-solvent interaction for large macromolecules is very complicated. It is often useful to refer to results from model compounds to understand solvation effects (c.f.

Section 5). However, because very few data are available concerning the compressibility of model compounds, we cannot base our answer on this kind of information. Instead, we can only make a qualitative argument. The total compressibility change for RNase A denaturation is approximately 0.01 mL mol^{-1} bar^{-1} (K. E. Prehoda, E. S. Mooberry, & J. L. Markley, manuscript in preparation). Because it is known from experiment that native RNase A molecule itself (excluding solvent effects) is compressible (Kundrot & Richards, 1987) and because the denatured RNase A molecule is expected to be much less compressible, the contribution to the total compressibility change from the protein itself should be negative. Therefore, to explain the experimental compressibility change, the contribution of solvation to the compressibility change must be positive. If the solvation layer itself is compressible, $\overline{K}_{int}^{\circ} > 0$, then this will follow from the fact that an unfolded protein presents a larger surface area to the solvent than a folded protein. This positive term must be larger than the negative term from the compressibility of the protein molecule itself (excluding solvation). This analysis implies that ΔV°_{obs} changes with pressure and that its accurate evaluation requires an analysis of compressibility effects.

We have assumed that the compressibility itself is independent of pressure. The linearity of pressure effects on NMR chemical shifts of folded proteins (Prehoda, 1997; Akasaka et al., 1997) is consistent with this hypothesis.

6. FUTURE PROSPECTS

The use of pressure as a variable for studying proteins is still largely unexplored. As discussed below, the volumetric consequences of several interesting protein properties may be usefully characterized through variable pressure experiments.

6.1. Pressure as a Tool for Probing Dynamic States of Proteins

On the basis of relative fluorescence intensities, it appears that pressure causes a transition from the native state of apomyoglobin (apoMb) to an intermediate state (Prehoda et al., 1996), one frequently described as a molten globule. The volume change for the native to intermediate transition at pH 5.8, 8 °C, is −85 ± 5 mL mol^{-1} (Prehoda et al., 1996). This volume change is very similar to those observed for the native to unfolded transition. A key question in the study of equilibrium molten globules is whether or not they are internally hydrated. If the intermediate state did not contain internal water, its partial molar volume would be much greater (~50%) than that of the native state (Ptitsyn, 1987). The nature of this state recently has been investigated by high pressure hydrogen exchange experiments as discussed in the following section.

6.2. High Pressure Hydrogen Exchange

The amide hydrogen exchange (HX) rate is a function of the local dynamics, structure and stability of a protein as described by the Linderstrøm-Lang model (Hvidt & Nielsen, 1966). HX rates at individual amides determined as a function of pressure from NMR data provide information on partial molar volume changes that occur during the local opening reactions as part of the exchange mechanisms and yield a detailed view of protein dynamics and stability at a local level. Pressure-dependent HX from the folded state will complement the approach developed by Englander for temperature-dependent HX from the folded state (Bai et al., 1995ab). The Englander experiment provides information on ΔG and m values (i.e., surface area exposed) for the local unfolding reactions. The analogous pressure experiment will report on ΔG and ΔV values. Complications of the first approach are thermal effects on structure (entropy effects); complications of the second are compressibility effects on structure. Both approaches yield information on cooperatively folded structural units.

Two different strategies can be used to investigate HX as a function of T and p: spectra can be acquired in the NMR probe at elevated pressures to follow exchange in real time, or exchange can be carried out in a pressure bomb outside the magnet, and aliquots can be examined at atmospheric pressure. The latter approach has the advantage of providing better resolution and signal-to-noise. The former procedure has the advantage of allowing faster rates to be examined and allows one to measure lifetimes of exchangeable protons by magnetization transfer from the solvent.

An HX experiment carried out on apoMb at 2.5 kbar, a pressure at which the protein appears to be 98% in the I state, revealed greatly reduced protection for sites in helix E of the N state, but continued protection for sites located in helices A, B, G, and H of the N state (K. E. Prehoda, S. N. Loh, B. F. Volkman, E. S. Mooberry, & J. L. Markley, unpublished). The observed protection patterns of apoMb at high pressure are consistent with those observed for the intermediate state at low pH and atmospheric pressure (Hughson & Baldwin, 1989).

REFERENCES

Akasaka, K., Tizuka, T., & Yamada, H. (1997) Pressure-Induced Changes in the Folded Structure of Lysozyme. *J. Mol. Biol. 271*, 671–678.

Attwood, P. V., & Gutfreund, H. (1980) The Application of Pressure Relaxation to the Study of the Equilibrium Between Metarhodopsin I and II From Bovine Retinas. *FEBS Lett. 119*, 323–326.

Bai, Y., Englander, J. J., Mayne, L., Milne, J. S., & Englander, S. W. (1995a) Thermodynamic Parameters from Hydrogen Exchange Measurements. *Methods Enzymol. 259*, 344–356.

Bai, Y., Sosnick, T. R., Mayne, L., & Englander, S. W. (1995b) Protein Folding Intermediates: Native-State Hydrogen Exchange. *Science 269*, 192–197.

Ballard, L., Reiner, C., & Jonas, J. (1996) High-Resolution NMR Probe for Experiments at High Pressures. *J Magn Res A 123*, 81–86.

Bai, Y., Sosnick, T. R., Mayne, L., & Englander, S. W. (1995) Protein Folding Entermediates: Native-State Hydrogen Exchange. *Science 269*, 192–7.

Benedek, G. B., & Purcell, E. M. (1954) Nuclear Magnetic Resonance in Liquids under High Pressure. *J. Chem. Phys. 22*, 2003–2012.

Brandts, J. F., Oliveira, R. J., & Westort, C. (1970) Thermodynamics of Protein Denaturation. Effect of Pressure on the Denaturation of Ribonuclease A. *Biochemistry 9*, 1038–1047.

Bridgman, P. W. (1914) Pressure Effect on Egg Albumin. *J. Biol. Chem. 19*, 511.

Bridgman, P. W. (1931) *The Physics of High Pressure*, Macmillan Co., New York.

Callen, H. B. (1985) *Thermodynamics and an Introduction to Thermostatistics*, 2nd Ed., John Wiley.

Chalikian, T. V., & Breslauer, K. J. (1996) On Volume Changes Accompanying Conformational Transitions of Biopolymers. *Biopolymers 39*, 619–626.

Coates, J. H., Criddle, A. H., & Geeves, M. A. (1985) Pressure-Relaxation Studies of Pyrene-Labelled Actin and Myosin Subfragment 1 from Rabbit Skeletal Muscle. *Biochem J. 232*, 351–356.

Connolly, M. L. (1985) Computation of Molecular Volume. *J. Am. Chem. Soc. 107*, 1118–1124.

Connolly, M. L. (1993) The Molecular Surface Package. *J. Mol. Graphics 11*, 139–141.

Da Poian, A., Oliveira, A., Gaspar, L., Silva, J., & Weber, G. (1993) Reversible Pressure Dissociation of R17 Bacteriophage. The Physical Individuality of Virus Particles. *J. Mol. Biol. 231*, 999–1008.

Davis, J. S. (1981) The Influence of Pressure on the Self-Assembly of the Thick Filament from the Myosin of Vertebrate Skeletal Muscle. *Biochem J. 197*, 301–308.

Davis, J. S. (1985) Kinetics and Thermodynamics of the Assembly of the Parallel- and Antiparallel-Packed Sections of Synthetic Thick Filaments of Skeletal Myosin: A Pressure-Jump Study. *Biochemistry 24*, 5263–5269.

Dill, K. (1990) Dominant Forces in Protein Folding. *Biochemistry 29*, 2357–2372.

Drickamer, H. G., Frank, C. W., & Slichter, C. P. (1972) Optical Versus Thermal Transitions in Solids at High Pressure. *Proc. Natl. Acad. Sci. 69*, 933–937.

Ehrhardt, M. R., Erijman, L., Weber, G., & Wand, A. J. (1996) Molecular Recognition by Calmodulin-Pressure-Induced Reorganization of a Novel Calmodulin-Peptide Complex. *Biochemistry 35*, 1599–1605.

Erijman, L., Paladini, A. A., Lorimer, G. H., & Weber, G. (1993) Plurality of Protein Conformations of Ribulose-1,5-Bisphosphate Carboxylase/Oxygenase Monomers Probed by High Pressure Electrophoresis. *J. Biol. Chem. 268*, 25914–25919.

Frey, U., Helm, L. & Merbach, A. E. (1990) High-Pressure, High Resolution NMR Probe Working at 400 MHz. *High Pressure Res. 2*, 237–245.

Frye, K. J., Perman, C. S., & Royer, C. A. (1996) Testing the Correlation between ΔA and ΔV of Protein Unfolding. *Biochemistry 35*, 10234–10239.

Frye, K. J., & Royer, C. A.. (1997) The Kinetic Basis for the Stabilization of Staphylococcal Nuclease by Xylose. *Protein Sci. 6*, 789–793.

Gill, S. J., & Glogovsky, R. L. (1965) Pressure Denaturation of Ribonuclease A. *J. Phys. Chem. 69*, 1515–1519.

Hamann, S. D. (1980) The Role of Electrostriction in High Pressure Chemistry. *The Review of Physical Chemistry of Japan 50*, 147–168.

Harrison, P. M., Bamborough, P., Daggett, V., Prusiner, S. B., & Cohen, F. E. (1997) The Prion Folding Problem. *Current Opinion in Structural Biology 7*, 53–9.

Hawley, S. A. (1971) Reversible Pressure-Temperature Denaturation of Chymotrypsinogen. *Biochemistry 10*, 2436–2442.

Heremans, K. (1982) High Pressure Effects on Proteins and other Biomolecules. *Ann Rev Biophys Bioeng 11*, 1–21.

Holzapfel, W. B., & Issacs, N, S. (Eds.) (1997) *High Pressure Techniques in Chemistry and Physics, a Practical Approach*. The Practical Approach in Chemistry Series, Oxford Univ. Press, Oxford, U.K.

Hughson, F. M. and Baldwin, R. L. (1989) Use of Site-Directed Mutagenesis to Destabilize Native Apomyoglobin Relative to Folding Intermediates. *Biochemistry 28*, 4415–4422.

Hvidt, A., & Nielsen, S. O. (1966) Hydrogen Exchange in Proteins. *Adv Protein Chem 21*, 287–386.

Jonas, J., Koziol, P., Peng, X., Reiner, C., & Campbell, D. M. (1993) High Resolution NMR Spectroscopy at High Pressures. *J. Magn. Reson., Ser. B 102*, 299–309.

King, L., & Weber, G. (1986) Conformational Drift of Dissociated Lactate Dehydrogenases. *Biochemistry 25*, 3632–3637.

Kornblatt, M. J., Kornblatt, J. A., & Hui Bon Hoa, G. (1993) The Role of Water in the Dissociation of Enolase, a Dimeric Enzyme. *Arch. Biochem. Biophys. 306*, 495–502.

Kundrot, C. E., & Richards, F. M. (1987) Crystal Structure of Hen Egg-White Lysozyme at a Hydrostatic Pressure of 1000 Atmospheres. *J. Mol. Biol. 193*, 157–170.

Li, T. M., Hook, J.W., Drickamer, H.G. & Weber, G. (1976) Plurality of Pressure-Denatured Forms in Chymotrypsinogen and Lysozyme. *Biochemistry 15*, 5571–5580.

Macgregor, R. B., Clegg, R. M., & Jovin, T. M. (1985) Pressure-Jump Study of the Kinetics of Ethidium Bromide Binding to DNA. *Biochemistry 24*, 5503–5510.

Makhatadze, G. I., Clore, G. M., & Gronenborn, A. M. (1995) Solvent Isotope Effect and Protein Stability. *Nature Struct. Bio. 2*, 852–855.

Makhatadze, G. I., & Privalov, P. L. (1993) Contribution of Hydration to Protein Folding Thermodynamics I. The Enthalpy of Hydration. *J. Mol. Biol. 232*, 639–659.

Markley, J. L. (1975) Correlation Proton Magnetic Resonance Studies at 250 MHz of Bovine Pancreatic Ribonuclease. I. Reinvestigation of the Histidine Peak Assignments. *Biochemistry 14*, 3546–54.

Markley, J. L., Royer, C. A., & Northrop, D., Eds. (1996) *High-Pressure Effects in Molecular Biophysics and Enzymology*, Oxford University Press, New York, 1996.

Masterton, W. L. (1954) Partial Molal Volumes of Hydrocarbons in Water Solution. *J. Chem. Phys. 22*, 1830.

Mooberry, E. S., Prehoda, K. E., & Markley (1996) High Pressure Double Resonance NMR Probe, 37th Experimental NMR Conference, Asilomar, CA, March 17–22.

Mooberry, E. S., Prehoda, K. E., & Markley (1997) High Pressure High Resolution Triple Resonance NMR Probe for Protein Studies, 38th Experimental NMR Conference, Orlando, FL, March 23–27.

Morild, E. (1981) The Theory of Pressure Effects on Enzymes. *Adv. Prot. Chem. 34*, 93–166.

Morishima, I., Ogawa, S., Yamada, H., (1980) High-Pressure Proton Nuclear Magnetic Resonance Studies of Hemoproteins. Pressure-Induced Structural Change in Heme Environments of Myoglobin, Hemoglobin and Horseradish Peroxidase. *Biochemistry 19*, 1569–1575.

Oliveira, A. C., Gaspar, A., Da Poian, T., & Silva, J. L. (1994) Arc Repressor Will Not Denature Under Pressure in the Absence of Water. *J. Mol. Biol. 240*, 184–190.

Pin, S., Royer, C. A., Gratton, E., Alpert, B., & Weber, G. (1990) Subunit Interactions in Hemoglobin Probed by Fluorescence and High-Pressure Techniques. *Biochemistry 29*, 9194–9201.

Prehoda, K. E. (1997) *Volumetric Analysis of Proteins by High Pressure Denaturation*, Ph.D. Thesis, University of Wisconsin-Madison.

Prehoda, K. E. & Markley, J. L. (1996) Use of Partial Molar Volumes of Model Compounds in the Interpretation of High-Pressure Effects on Proteins, In *High-Pressure Effects in Molecular Biophysics and Enzymology* (J. L. Markley, C. A. Royer, & D. Northrop, Eds.), Oxford University Press, pp. 33–43.

Prehoda, K. E., Loh, S. N., & Markley, J. L. (1996) Modeling Volume Changes in Proteins using Partial Molar Volumes of Model Compounds, In *Techniques in Protein Chemistry VII* (D. R. Marshak, Ed.), Academic Press, New York, pp. 433–438.

Ptitsyn, O. B. (1987) Protein folding: Hypotheses and Experiments. *J. Protein Chem. 6*, 273–293.

Richards, F. M. (1977) Areas, Volumes, Packing, and Protein Structure. *Ann. Rev. Biophys. Bioeng. 6*, 151–176.

Rodgers, K., Pochapsky, T., & Sligar, S. (1988) Probing the Mechanisms of Macromolecular Recognition: the Cytochrome b_5-Cytochrome c Complex. *Science 240*, 1657–1659.

Royer, C. A. (1995) Application of Pressure to Biochemical Equilibria: The Other Thermodynamic Variable. *Methods in Enzymology 259*, 357–377.

Royer, C. A., Weber, G., Daly, T., & Matthews, K. (1986) Dissociation of the Lactose Repressor Protein Tetramer using High Hydrostatic Pressure. *Biochemistry 25*, 8308–8315.

Royer, C. A., Hinck, A. P., Loh, S. N., Prehoda, K. E., Peng, X., Jonas, J., & Markley, J. L. (1993) Effects of Amino Acid Substitutions on the Pressure Denaturation of Staphylococcal Nuclease As Monitored by Fluorescence and NMR Spectroscopy. *Biochemistry 32*, 5222–5232.

Ruan, K., & Weber, G. (1988) Dissociation of Yeast Hexokinase by Hydrostatic Pressure. *Biochemistry 27*, 3295–3303.

Ruan, K., & Weber, G. (1989) Hysteresis and Conformational Drift of Pressure-Dissociated Glyceraldehydephosphate Dehydrogenase. *Biochemistry 28*, 2144–2153.

Ruan, K., & Weber, G. (1993) Physical Heterogeneity of Muscle Glycogen Phosphorylase Revealed by Hydrostatic Pressure Dissociation. *Biochemistry 32*, 6295–6301.

Samarasinghe, S. D., Campbell, D. M., Jonas, A., & Jonas, J. (1992) High-Resolution NMR Study of the Pressure-Induced Unfolding of Lysozyme. *Biochemistry 31*, 7773–7778.

Sawamura, S., Kitamura, K., & Taniguchi, Y. (1989) Effect of Pressure on the Solubilities of Benzene and Alkylbenzenes in Water. *J. Phys. Chem. 93*, 4931–4935.

Shahidi. (1981) Partial Molar Volumes of Organic Compounds in Water. Part 8-Benzene Derivatives. *J. Chem. Soc., Faraday Trans. 77*, 1511–1514.

Shahidi, F., Farrel, P. G., & Edward, J. T. (1977) Partial Molar Volumes of Organic Compounds in Water. Part 2-Amines and Amides. *J. Chem. Soc., Faraday Trans. 73*, 715.

Silva, J. L., Miles, E. W., & Weber, G. (1986) Pressure Dissociation and Conformational Drift of the Beta Dimer of Tryptophan Synthase. *Biochemistry 25*, 5780.

Spolar, R. S., Livingstone, J. R., & Record, M. T. (1992) Use of Liquid Hydrocarbon and Amide Transfer Data To Estimate Contributions to Thermodynamic Functions of Protein Folding from the Removal of Nonpolar and Polar Surface from Water. *Biochemistry 31*, 3947–3955.

Tamura, Y., & Gekko, K. (1995) Compactness of Thermally and Chemically Denatured Ribonuclease A As Revealed by Volume and Compressibility. *Biochemistry 34*, 1878–1884.

Vidugiris, G. J. A. Markley, J. L. and Royer, C. A. (1995) Evidence for a Molten Globule-Like Transition State in Protein Folding from Determination of Activation Volumes. *Biochemistry 34*, 4909–4912.

Vidugiris, G. J., Trucksess, D. M., Markley, J. L., & Royer, C. A. (1996) High-Pressure Denaturation of Staphylococcal Nuclease Proline-to-Glycine Substitution Mutants. *Biochemistry 35*, 3857–3864.

Visser, A., Li, T., Drickamer, H.G. & Weber, G. (1977) Volume Changes in the Formation of Internal Complexes of Flavinyltryptophan Peptides. *Biochemistry 16*, 4883–4886.

Wagner, G. (1980) Activation Volumes for the Rotational Motion of Interior Aromatic Rings in Globular Proteins Determined by High Resolution ^{1}H NMR at Variable Pressure. *FEBS Letters, 112*, 280–284.

Weber, G., & Drickamer, H. G. (1983) The Effect of High Pressure upon Proteins and Other Biomolecules. *Quart. Rev. Biophys. 16*, 89–112.

Yonker, C. R., Zemanian, T. S., Wallen, S. L., Linehan, J. C. & Franz, J. A. (1995) A New Approach for the Convenient Measurement of NMR Spectra in High-Pressure Liquids. *J. Magn. Reson. Ser. A. 113*, 102–107.

Zahl, A., Neubrand, A., Aygen, S., & van Eldik, R. (1994) A High Pressure NMR Probehead for Measurements at 400 MHz. *Rev. Sci. Instrum. 65*, 882–886.

Zhang, J., Peng, X., Jonas, A. & Jonas J. (1995) NMR Study of the Cold, Heat, and Pressure Unfolding of Ribonuclease A. *Biochemistry 34*, 8631–8641.

Zipp, A., & Kauzmann, W. (1973) Pressure Denaturation of Metmyoglobin. *Biochemistry 12*, 4217–4228.

6

THE ROLE OF MOLECULAR RECOGNITION IN ACTIVATION AND REGULATION IN THE GROWTH HORMONE-PROLACTIN FAMILY OF HORMONES AND RECEPTORS

Anthony A. Kossiakoff,* Celia Schiffer, and Abraham M. de Vos

Department of Protein Engineering
Genentech, Inc.
460 Point San Bruno Blvd.
South San Francisco, California 94080

The mechanism by which cells communicate is one of the most intensely studied problems in biology. This communication is generally thought to be mediated through cell-surface receptors, initiated by a ligand binding event at the extracellular portion of the receptor. The details at the molecular level of how this extracellular message is transduced through the membrane to the interior of the cell are only now beginning to emerge. In this paper we discuss some of the work currently ongoing in this lab aimed at delineating the structural basis for the signal transduction process.

We have focused on the endrocrine family of human hormones and receptors. These include the hormones: growth hormone, placental lactogen and prolactin; and the growth hormone and prolactin receptors. The sequence homology of the extracellular domains (ECDs) of these receptors classifies them as belonging to the hematopoietic receptor superfamily (Bazan, 1990). They have a three-domain organization: the extracellular portion that binds the activating hormone; a transmembrane segment of about 25 amino acids; and a cytoplasmic domain. Biological response is triggered by hormone binding that drives receptor aggregation (Cunningham et al.,1991, De Vos et al., 1992). The aggregated receptor complex binds one or several tyrosine kinases belonging to the JAK family, which then trans-phosphorylate elements on themselves, the receptors and associated transcription factors (Ihle et al., 1994).

The structure of the human growth hormone (hGH) bound to its receptor (hGH$_R$) first showed how the mechanism of receptor aggregation worked (De Vos et al., 1992).

* Telephone: (773) 834-2846.

Protein Dynamics, Function, and Design, edited by Jardetzky *et al.*
Plenum Press, New York, 1998

One hormone binds two receptors. This finding initially presented a puzzle; how do two identical receptor ECDs bind to one hormone considering that hGH has no semblance of symmetry that could support binding of each of the ECDs in a similar fashion. (hGH is an asymmetric four-helix bundle protein). Surprisingly, the high resolution structure revealed that the same regions of each individual receptor bind to two topographically very different regions on hGH. This is illustrated in Figure 1, which shows a molecular surface rendering of hGH. At binding Site 1, hGH has a large invagination into which the contacting side chains of one of the receptor ECDs insert. The surface area that becomes buried on the hormone is about 1300 Å^2. Binding Site 2 is much flatter and the surfaces, both of the hGH and the receptor (ECD2), appear not to be as complementary. In comparison, Site 2 buries only 850 Å^2 of hGH's surface. The difference in surface interactions presumably plays an active physiological role. It has been shown that the 2 ECD:1 hGH complex is formed in a stepwise manner; binding Site 1 is a high affinity site and is always occupied before binding Site 2 (Fuh et al., 1992).

The binding of the endocrine hormones to their respective receptors is a highly regulated process. hGH has the ability to activate the prolactin receptor (hPRL$_R$), whereas pro-

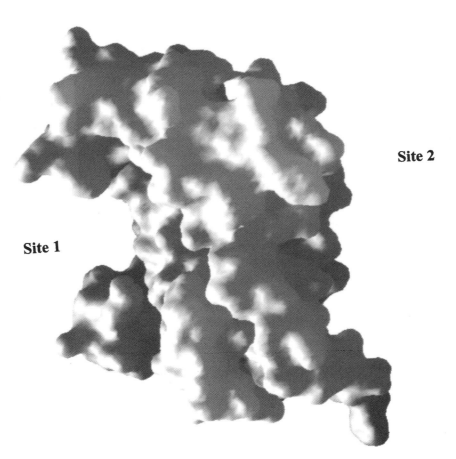

Figure 1. Solvent accessible surface of hGH showing the topographically different Site 1 and Site 2 binding sites.

lactin (hPRL) does not bind to the growth hormone receptor. Human placental lactogen (hPL) activates hPRL$_R$, although the hPRL and hPL have only limited homology (25%) (Nicoll et al., 1986). In contrast, hPL binds 2300 fold weaker than does hGH to hGH$_R$, even though they have 85% sequence identity (Lowman, et al., 1991). hPRL and hPL can be recruited to bind to hGH$_R$ by altering only about half a dozen critical residues (Lowman et al., 1991; Cunningham et al. 1990). Growth hormones are highly conserved, and hGH can activate GH$_R$s of other species; surprisingly, the converse is not true- the GHs of other species cannot activate the human receptor (Boutin et al., 1989).

Taken together, these observations paint a puzzling picture. On the one hand, formation of the active 2:1 complex requires the GH$_R$s and PRL$_R$s to be versatile enough to bind two topographically highly diverse surfaces on the hormone. On the other, specificity is apparently dictated by relatively few key residues, because molecules very similar to hGH do not bind to hGH$_R$.

THE STRUCTURE OF HUMAN PLACENTAL LACTOGEN

We have recently solved the structure of hPL to 2.2 Å resolution (Kossiakoff, Somers, De Vos, unpublished). In most respects the structure of this hormone closely mirrors that of hGH. With this structure in hand, comparing the structures of hGH bound to hGH$_R$ and the hPRL$_R$ indicates that several changes occur in the hormone between the two complexes. The most pronounced of these changes are in the mini-helix 1 region and the loop connecting helices 1 and 2 (Figure 2). Noting the conformation of the mini-helix, it is apparent that there is striking structural similarity between free hPL and hGH when it is bound to the hPRL$_R$. In fact, in these two structures the mini-helix is melted out into a new type of loop structure. It is tempting to think, based on the structures, that changes in the mini-helix and associated loop conformations are the principal determining factor in hPL's specificity toward the prolactin receptor.

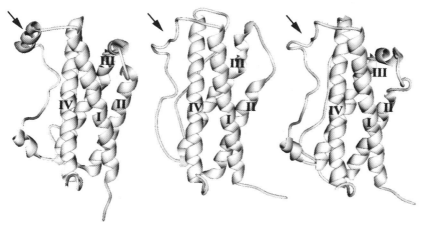

Figure 2. Ribbon diagrams of: left panel–hGH when bound to the hGH receptor, middle–hPL and, right panel–hGH when bound to the prolactin receptor. The position of the "mini-helix" is shown by the arrow.

THE STRUCTURE OF THE HGH "SUPERMUTANT"-RECEPTOR COMPLEX

A variant of hGH that binds 400 times tighter to its receptor was produced using a technique called "phage display". This method has been described in detail in some recent reviews (Wells and Lowman, 1992) and so it will not be discussed here. In outline, a protein of choice (in this case hGH) can be expressed on the surface of bacteriophage M13. M13 is the phage that is commonly used in *E coli* mutagenesis, and the gene for hGH has been inserted into the phage genome, where it can be easily mutated by standard techniques. Specific residues can be targeted for random mutagenesis; for hGH 15 sites from 5 different libraries were mutated (Lowman et al., 1991; Lowman and Wells, 1993). Each library had about 3×10^6 random mutants and although not all combinations of the 20 amino acids at each site were sampled, the results certainly included a very extensive subset. The choice of which residues to mutate was made based on a previous alanine-scan mutagenesis (Cunningham and Wells, 1989; Cunningham and Wells, 1993). That analysis identified those residues that most influenced hormone binding; these residues were generally found on helix 4 and the loop connecting helix 1 and helix 2.

Figure 3 shows a ribbon diagram of hGH on which are marked the sites that were mutated during the phage display selections (Ultsch et al., 1994). Several characteristics are noteworthy. Three mutations were located in the Zn^{++} binding site: H18D, H21N and E174S. The side chains of H18 and E174 act as ligands in the hGH-prolactin complex and H21 plays a supporting role in the interaction by forming an H-bond to the carboxylate of E174 (Somers et al., 1994; Kossiakoff et al., 1994). The similar interaction in the hGH-hGHR does not require

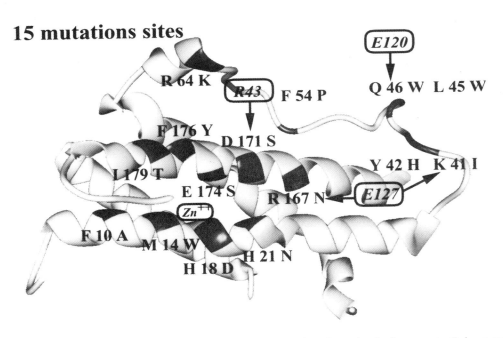

Figure 3. Ribbon representation of hGH with the positions of the sites of mutation for "super-mutant" shown as strips and labeled. Circled residues are hGH receptor amino acids that make H-bonds in the hGH-hGH_R complex to the wild-type hGH residues.

Zn^{++} binding and, in fact, mutagenesis data show that the coordinating side chains of H18, H21, and E174 have a negative influence on binding affinity in this latter system (Cunningham and Wells, 1993). Thus, it is not surprising that these residues were significantly altered in their character in the phage selections. Another point of interest is the finding that a few important changes that increased hormone affinity were residues whose side chains were actually buried and did not make direct contact with the receptor itself: F10A and M14W. These are what are called "framework" residues, residues that alter the local scaffolding of the chain in such a way that presentation of neighboring side chains is affected slightly differently compared to the wild-type arrangements. The effect of framework residues has been appreciated in antibody engineering where it is known that non-CDR residues can produce appreciable effects through indirect packing effects (Eigenbrot et al., 1993).

An accepted tenet of protein-protein interactions is that ionic interactions have a positive effect on binding energies. The ala-scan mutagenesis generally supported this notion (Cunningham and Wells, 1993). It was, therefore, very surprising that so many H-bonding interactions were eliminated in the phage selection mutagenesis: D171- receptor R43 (D replace with S), R167- receptor E127 (R to N), Q46- receptor E120 (Q to W) and K41-receptor E127 (K to I). K41 and Q46 are contained in a segment of chain referred to as Mini-helix 1. The residues that interfaced the receptor (residues K41,Y42,L45 and Q46) were mutated together in one library. The clone that displayed the strongest binding (4.5 times wild-type) had a sequence of I, H, W, W at these positions. Residues 41 and 46 change from hydrophilic to hydrophobic and 42, the opposite. Taken in the context of the sequence alone this is a puzzling result: how can one surface of an interface change stereochemical character to such a degree and remain complementary to the other surface?

The above situation is indeed a conundrum if the phage display mutagenesis is viewed simply as a procedure by which different amino acid side chains are displayed on a conformationally invariant main chain backbone. Such is not the case, as is shown in Figure 4. In this figure three conformational states of mini-helix 1 of hGH are shown. The first state is the helix in the context of the 2:1 complex (panel a). Panel (c) shows the helix in the free supermutant structure, and the last panel (b) is the mini-helix conformation as it exists in the supermutant complex (Schiffer, Kossiakoff, unpublished). It is apparent that for both the hGH bound to its receptor and for the uncomplexed supermutant (panels a and c) this segment of chain remains formed as a helix. However, for the supermutant when it is complexed to the receptor, the mini-helix has melted out, giving a more extended structure. The side chains of the mutated groups are disposed in quite different regions of space compared to where they would be if the segment of chain were to remain a helix. There is a very important lesson in this observation. Phage display mutagenesis not only probes sequence diversity, but in some types of tertiary structures conformational diversity, as well. Thus, conformational space was probed in the case of the mini-helix and framework residues as a convolution of these two parameters making the actual diversity available much larger than the computed 3×10^6 that was based on amino acid composition alone.

CONCLUSIONS

Within this framework of hormone-receptor interactions there is an extraordinary adaptability among these molecules to synthesize binding epitopes to a wide variety of binding interfaces. The nature of the adjustments required to form the optimum binding surfaces suggests that recognition and binding is directed by an induced-fit mechanism. Distinct from the process of molecular recognition associated with the antibody-antigen paradigm, where

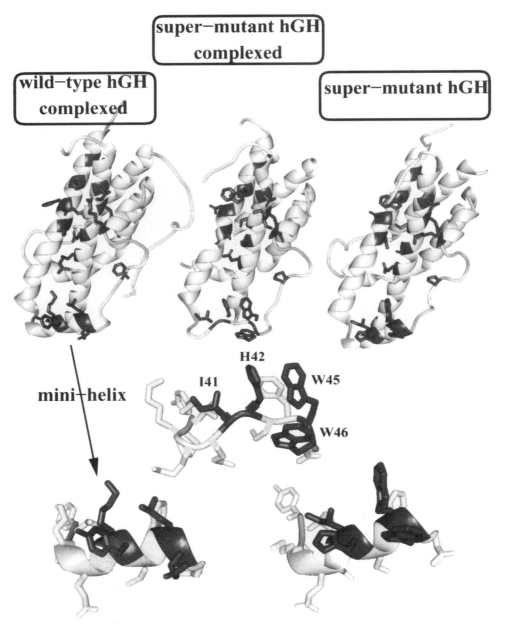

Figure 4. Changes in the mini-helix conformation of: left) hGH complexed to hGHR, center) super-mutant complexed to hGHR, and right) free super-mutant. Note that for the complexed super-mutant the helix is melted out. Mutated residues on the mini-helix are shown.

binding is developed through sequence diversity, the endocrine receptors use essentially a constant set of residues to bind surfaces that are diverse both in sequence and in conformation. This is accomplished by employing conformational diversity, both local and global. This theme is most likely not limited to cytokine- receptor systems and other systems will be found that use this strategy to develop their binding and specificity determinants.

REFERENCES

Bazan, J. F. 1990. Structural design and molecular evolution of a cytokine receptor superfamily. Proc. Natl. Acad. Sci. U.S.A. 87:6934–6938.

Boutin, J. M., M. Edrey, M. Shirota and et al. 1989. Identification of a cDNA encoding a long form of prolactin receptor in human hepatoma and breast cancer cells. Mol. Endocrinol. 3:1455–1466.

Cunningham, B. C., S. Bass, G. Fuh and J. A. Wells. 1990. Zinc mediation of the binding of human growth hormone to the human prolactin receptor. Science 250:1709–1712.

Cunningham, B. C., M. Ultsch, A. M. De Vos, M. G. Mulkerrin, K. R. Clauser and J. A. Wells. 1991. Dimerization of the extracellular domain of the human growth hormone receptor by a single hormone molecule. Science 254:821–825.

Cunningham, B. C. and J. A. Wells. 1989. High-resolution epitope mapping of hGH-receptor interactions by alanine-scanning mutagenesis. Science 244: 1081–1085.

Cunningham, B. C. and J. A. Wells. 1993. Comparison of a structural and a functional epitope. J. Mol. Biol. 234:554–563.

De Vos, A. M., M. Ultsch and A. A. Kossiakoff. 1992. Human Growth Hormone and Extracellular Domain of its Receptor: Crystal Structure of the Complex. Science 255:306–312.

Eigenbrot, C., M. Randal, L. Presta, P. Carter and A. A. Kossiakoff. 1993. X-ray Structures of the Antigen-Binding Domains from Three Variants of Humanized anti-p185HER2 Antibody 4D5 and Comparison with Molecular Modelling. J. Mol. Biol. 229:969–995.

Fuh, G., B. C. Cunningham, R. Fukunaga, S. Nagata, D. V. Goeddel and J. A. Wells. 1992. Science 256: 1677–1680.

Ihle, J. N., B. A. Witthuhn, F. W. Quelle, K. Yamamoto, W. E. Thierfelder, B. Kreider and O. Silvennoinen 1994. Signaling by the cytokine receptor superfamily: JAKs and STATs. Trends Biochem. Sci. 19:222–227.

Kossiakoff, A. A., W. Somers, M. Ultsch, K. Andow, Y. Muller, and A. M. De Vos. 1994. Comparison of the intermediate complexes of human growth hormone bound to the human growth hormone and prolactin receptors. Prot. Sci. 3:1697–1705.

Lowman, H. B., S. H. Bass, N. Simpson and J. A. Wells. 1991. Selecting high-affinity binding proteins by monovalent phage display. Biochemistry 30:10832–10838.

Lowman, H. B., B. C. Cunningham and J. A. Wells. 1991. Mutational analysis and protein engineering of receptor-binding determinants in human placental lactogen. J. Biol. Chem. 266:10982–10988.

Lowman, H. B. and J. A. Wells. 1993. Affinity maturation of human growth hormone: Monovalent phage display. J. Mol. Biol. 234:564–578.

Nicoll, C. S., G. L. Mayer and S. M. Russel. 1986. Structural features of prolactins and growth hormones that can be related to their biological properties. Endocrine Rev. 7:169–203.

Somers, W., M. Ultsch, A. M. De Vos and A. A. Kossiakoff. 1994. The X-ray structure of the growth hormone-prolactin receptor complex: Receptor binding specificity developed through conformational variability. Nature 372:1697–1705.

Ultsch, M., W. Somers, A. A. Kossiakoff and A. M. De Vos. 1994. The crystal structure of affinity-matured human growth hormone at 2 Å resolution. J. Mol. Biol. 236:286–299.

Wells, J. A. and H. B. Lowman. 1992. Rapid evolution of peptide and protein binding properties in vitro. Curr. Opin. Struct. Biol. 2:597–604.

7

PROTEIN DYNAMICS AND FUNCTION

Hans Frauenfelder

Center for Nonlinear Studies
Los Alamos National Laboratory
Los Alamos, New Mexico 87545

1. INTRODUCTION

Proteins appear at the same time to be rather simple and extremely complex. The overall reaction of a given protein can often be described in a very simple equation, but as soon as one delves more deeply into the mechanism of the reaction, it becomes more and more sophisticated. Proteins thus can be understood on a number of different levels. Ultimately the goal is to have predictive power and to know how a given mutation, for instance, changes the function. We are still far away from this goal, but we can outline some of the steps needed to arrive at the goal: The structure, the energy landscape, and the dynamics of the particular protein have to be known and understood and the laws govern the reactions have to be in hand. In the following, some of the essential steps will be sketched. Since a complete description of the various steps is not possible here, enough references are given.

2. REACTIONS

Most protein functions involve chemical reactions of some form, and an understanding of the theory underlying such reactions is essential. Most current texts short-change the modern treatment of reaction. A brief sketch of some of the essential features may consequently be useful (Frauenfelder and Wolynes, 1985; Hänggi et al, 1990; Fleming and Hänggi, 1993). Usually, a chemical reaction is described as corresponding to an event in which an enthalpy barrier of height H has to be overcome and the reaction rate coefficient $k(T)$ at the temperature T is then given by an Arrhenius equation,

$$k(T) = A (T/T_o) \exp(-H/RT). \tag{1}$$

Here R is the gas constant, A an empirical preexponential factor, and T_o a reference temperature (100K). The experimental data on protein reactions are usually taken over a small range in temperatures and Eq (1) can always be used, but it does not do justice to the actual science. To realize the complexity in an actual protein reaction, consider the "simple" binding of carbon monoxide to myoglobin:

$$Mb + CO \leftrightarrow MbCO. \tag{2}$$

Initially it was assumed that this reaction is indeed just a one-step process, a transition over a fixed static barrier described by naive transition state theory. Detailed studies show, however, that a number of complicating factors occur. A few of these are important for understanding biological reactions:

1. *The Kramers Equation* (Kramers, 1940; Frauenfelder and Wolynes, 1985; Hänggi et al. 1990; Fleming and Hänggi, 1993). The Arrhenius relation, Eq. (1), describes a transition over a fixed barrier and it is assumed that the reacting system is always in equilibrium with its surrounding. Reality is different, however, and in 1940 Kramers showed that the reaction rate depends crucially on the coupling of the reacting system with the environment. In a useful approximation, this coupling can be described by the viscosity, η, of the medium surrounding the reaction partners. In the overdamped case, where the viscosity is large, the system "diffuses" over the barrier and the reaction rate decreases with increasing viscosity, $k(\eta) \approx c/\eta$. In the underdamped case, the transition is limited by the rate at which the system acquires enough energy to move over the barrier and it is proportional to the viscosity, $k(\eta) \approx c\,\eta$. A living biological system is usually well thermostated and can change the viscosity more easily, for instance by a change in cholesterol, than the temperature. It can therefore adjust a reaction rate more easily by a change of viscosity than temperature. In the case of the binding reaction to myoglobin, Eq (2), a viscosity dependence has been found (Beece et al., 1980; Ansari et al., 1994; Kleinert et al., in press), suggesting that the simple form Eq (1) is not adequate.

2. *Nonadiabatic Reactions* (Landau, 1932; Zener, 1932; Stückelberg, 1932; Frauenfelder and Wolynes, 1985; Hänggi et al., 1990). In chemical reactions, both electrons and nuclei move. Often, the electrons follow the nuclear motion adiabatically and the rate coefficient for a transition can be treated as indicated above. If, however, the electrons adjust only slowly to the changing nuclear coordinates, the reaction becomes non-adiabatic and the rate coefficient is reduced below the adiabatic value given, for instance, by the Kramers equation.

3. *The Tunnel Effect* (Hänggi et al., 1990). At low temperatures, quantum effects can become important. Because of their small mass, electrons tunnel easily. But even protons, and somewhat surprisingly, molecules as heavy as CO can tunnel (Alberding et al., 1976). It has even been possible to investigate the isotope effect (Alben et al., 1980). While molecular tunneling most likely does not play a role in biological phenomena, its properties can be used to investigate the details of biological reactions.

4. *Fluctuating Barriers* (Stein et al., 1989). The usual treatment of a reaction or the crossing of an enthalpy barrier assumes that the barrier is fixed. In biological systems, however, the height of the barrier may fluctuate and these fluctuations must be taken into account in the treatment of the reaction. Even resonance phenomena may occur (Doering and Gadoua, 1992).

5. *Non-Arrhenius Temperature Dependence.* In complex systems, such as glasses and proteins, the temperature dependence of the rate coefficient usually does not

follow an Arrhenius law, Eq (1). The data can, however, often be fitted by a Vo-gel-Fulcher relation,

$$k(T) = A \exp[-(H - H_o)/RT], \qquad (3)$$

or by a Ferry relation,

$$k(T) = A' \exp[-(E/RT)^2]. \qquad (4)$$

The appearance of a non-Arrhenius temperature dependence indicates that the reaction involves not just a small rigid part of the biological system, but that re-actions can involve a major part of the protein.

6. *Nonexponential Time Dependence.* The time course of first-order reactions, as in Eq. (1), is normally exponential. In proteins, deviations from exponentiality oc-cur (Austin et al., 1975). Nonexponential time dependence can have two causes. In the homogeneous case, each protein displays a nonexponential time depend-ence which can then usually be approximated by a stretched exponential,

$$N(t) = \exp[-(kt)^{\beta}], \qquad (5)$$

where the coefficient β is less than one. In the inhomogeneous case, each pro-tein reacts exponentially in time, but with a different rate coefficient. In this situation, the time dependence can be written as

$$N(t) = \int g(H) \exp[-k(H)t] \, dH \qquad (6)$$

where g(H) is the probability of finding a barrier between H and H + dH.

3. STRUCTURE AND ENERGY LANDSCAPE

Understanding, not just describing, proteins is the goal of protein studies. The im-portance of understanding is easily realized by considering the number of possible pro-teins. Assume that we construct all proteins consisting of about 200 amino acids and make one copy each. The result would fill about 10^{100} universa. Obviously, a systematic study of all possible proteins is impossible; we must understand their structure, dynamics, and function to be able to design new ones from first principles. We are still far away from this goal, but some signposts suggest how to proceed. Heisenberg wrote in 1972: "The history of physics.. is also a history of concepts. For an understanding of the phenomena, the first condition is the introduction of adequate concepts." If we look at the some of the main fields in physics, atoms, nuclei, particles, and solids, we notice that major progress has al-ways involved exploring and elucidating concepts in three major areas, structure, energy levels, and dynamics. Look at atomic physics: Major progress in understanding the struc-ture is related to the names of Rutherford and Bohr, the energy levels recall the names of Balmer and Bohr, the dynamics those of Heisenberg and Schrödinger. Are similar steps possible in the biomolecular physics and are some concepts already apparent?

The structure of a large number proteins is by now well known and the approach, based on X-ray diffraction and NMR, is discussed in related lectures. Three remarks are, however, relevant in the context of the present discussion. The first concerns the accuracy of the structure determination. It is not yet clear to which accuracy the active center of a protein has to be built to perform its function fully. It is consequently also not yet established if the present techniques give a sufficiently complete structure. The second remark concerns the distribution of structures. As will be discussed later, a given primary sequence can fold into a very large number of tertiary structures. The present structure determination techniques cannot give detailed information on all the structures that are present. Some of the most important ones may be missed. The third remark concerns intermediate structures that participate in the function. While some progress in the determination of the structure of such states has been made, not enough is known yet to fully describe even the simplest protein reaction.

The essential features of the structure of *Myoglobin* are well known (Dickerson and Geis, 1983; Stryer, 1995): Eight alpha helices form a box that contains a heme group with a central iron atom. Small ligands such as CO and O_2 bind reversibly at the iron atom. Overall, the reaction is as given in Eq. (l). This apparently very simple reaction turns out to be very sophisticated and even after many years of work it is not fully understood. The studies have, however, led to some crucial concepts that may be valid for many or all biomolecules.

One central concept that emerged is that of the *energy landscape* (Frauenfelder et al., 1991; Frauenfelder et al., 1997). A "simple" physical systems, such as an atom or a nucleus, in its ground state has a well defined energy and a well-defined structure. Its excited states can be characterized as energy levels. Standard pictures of proteins in textbooks give a similar impression: Each atom appears to be in its proper and unique place! This impression is completely misleading. As indicated in Figure 1, a given primary sequence can fold into a very large number of somewhat different structures, called conformational substates (CS), with essentially the same energy (Austin et al., 1975). Each CS is a valley in a conformational space of 3N dimensions, where N is the number of atoms in the protein. Figure 1 gives a highly schematized one-dimensional cross section through the energy landscape.

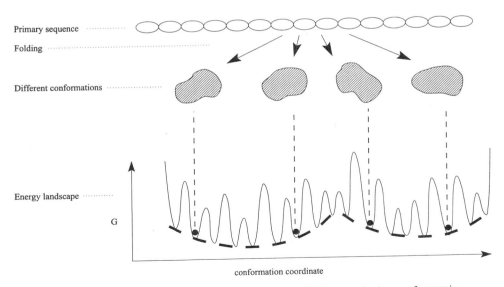

Figure 1. One-dimensional cross section through the simplified energy landscape of a protein.

The first evidence for the existence of conformational substates came from flash photolysis experiments with myoglobin (Austin et al., 1975). MbCO and MbO, can be photolyzed by a short light pulse. At low temperatures, below about 200K, the photodissociated ligand does not leave the protein, but rebinds geminately (internally). Rebinding is not exponential in time, as initially expected, but can be approximated by a power law. Using Eq. (6), it can be characterized by a distribution function g(H): Different myoglobin molecules, with identical primary sequences, possess different barriers for the binding of CO and O_2. The interpretation in terms of Figure 1 is straightforward. Myoglobin molecules in different substates possess slightly different structures which in turn result in different activation barriers H.

The existence of conformational substates and an energy landscape has been verified by many different approaches. Particularly striking are hole burning experiments (Friedrich, 1995). The principle of this method is easy to describe. Different substates can have slightly different energy levels. Spectral lines in different CS consequently appear at different wave numbers and a given spectral line is inhomogeneously broadened. By using a narrow laser line, a hole can be burned into such a broadened spectral band and the existence of the hole proves and explores the underlying energy landscape.

Further support for the concept of substates comes from a careful look at X-ray diffraction results (Frauenfelder et al., 1979). If CS have slightly different structures, a given atom should occupy somewhat different positions in different proteins. Such a distribution leads to a decrease in the intensity of the diffraction spots, given by the Debye-Waller factor. From the observed intensities, a measure of the spread in spatial distribution can be calculated. The experiments clearly show the spatial extent of the substates (Petsko and Ringe, 1984).

Figure 1 actually gives an oversimplified picture of the energy landscape. Comparing results from a broad range of experiments indicates that the energy landscape is arranged in a *hierarchy* (Ansari et al., 1985). A given primary sequence may fold into a small number of taxonomic substates. Taxonomic, because their properties, such as their average structure, enthalpy, entropy, and reaction rates can be described individually. Each of these taxonomic substates, substates of tier 0, can exist in a very large number of substates of tier 1. Each of these substates can in turn exist in a number of substates of tier 2, and so on. The average barrier between substates decreases with increasing tier numbers. At present, the organization and the details of the energy landscape are not fully known or understood even in one protein. Much work remains to be done!

4. DYNAMICS

Proteins are not static structures. Motions are crucial for the function of most, if not all, proteins. Take again myoglobin as an example. CO and O_2 enter myoglobin as if there was no protein present, but the Xray structure shows no cleft or channel leading to the active center. The protein consequently must undergo motions to open transient passageways. Similar observations have been made for other proteins. Thus, in order to understand and characterize the function of proteins, it is necessary to explore and characterize their dynamics.

Three different types of motions can be described with Figure 1. In *vibrations,* the protein remains within a given CS. *Fluctuations* occur when the protein jumps between different CS. These motions occur even of the protein is in thermal equilibrium. In *relaxation processes,* the protein is brought to a nonequilibrium state and it then returns to equilibrium by moving to CS with lower and lower Gibbs energies.

Fluctuations and relaxations are important for protein function. They are essentially classical phenomena and their rate coefficients usually depend strongly on temperature. While it is common to use the Arrhenius equation, Eq. (1) for their characterization, measurements over broader ranges in temperature demand often use of Eqs. (3) or (4). In either case, however, the rate coefficients decrease with decreasing temperature. Below some temperature, T_g, fluctuations and relaxations are so slow that the protein becomes effectively rigid.

Proteins share with glasses and spin glasses the existence of a rugged energy landscape and metastability below the glass transition (Iben et al., 1989). There are, however, a number of differences. The hierarchy of CS is more extended in proteins than in glasses and some of the motions of the protein depend on the surrounding. The main glass-like transition in proteins is *slaved* to the environment (Austin et al., 1975; Hagen et al., 1996; Kleinert et al., in press). This observation again shows that proteins cannot be treated as isolated systems, but that the interaction with the environment in the living system may be crucial for control.

5. FUNCTION

One of the ultimate goals of biophysical studies is to describe and understand the function of biomolecules in complete detail, combining knowledge from structure and dynamics investigations. We are not yet close to this goal in even one protein. Myoglobin may be the one case where this goal may be reachable. Even here, however, problems remain unsolved despite the fact that the number of papers published that treat the apparently simple reaction Eq. (1) is extremely large. Some of the general features and some of the problems will be sketched here.

The general features of the binding of small ligands such as CO and O_2 are shown in Figure 2. The ligand enters a pocket at the heme group and then binds covalently at the heme iron. The entire process looks deceptively simple, but becomes more sophisticated as one digs deeper. To explore the various phenomena, one starts with the bound state, for instance MbCO. A short laser flash breaks the Fe-CO bond and the CO moves into the pocket (Austin et al., 1975). At temperatures below about 200K, the CO does not leave the protein, but rebinds from within. This rebinding, as discussed earlier, gives information about the energy landscape and conformational substates. Above about 200K, the CO either rebinds or moves into the solvent. Measurements over a large range in time (ps to s) and temperatures (4–300K) suggest that the binding involves at least two separate steps, but raise a number of troubling questions of which only three are listed here: (i) CO enters the pocket or cavity in Mb and in protoheme equally fast, but leaves the Mb pocket about a thousand times slower in Mb. (ii) Below about 20K, the geminate binding occurs via quantum mechanical tunneling. Surprisingly, $^{12}C^{18}O$ tunnels faster than $^{13}C^{16}O$ even though it is heavier (Alben et al., 1980). (iii) In deoxyMb, a water molecule sits in the heme pocket (Quillin et al., 1995). Are these three observations related and how can they

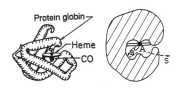

Figure 2. Left: The myoglobin skeleton. Right: Small molecules such as CO and O_2 enter a pocket at the heme group and covalently bind at the heme iron.

be integrated into a model with predictive power? A preliminary answer to these (and some other) questions can be given (Frauenfelder et al., in press) that involves the following points: The CO molecule, after entering the heme pocket, binds weakly to the pocket, with an enthalpy of about 20 kJ/mol. The exit from the pocket is consequently slowed down. Binding to the heme iron involves, at least at low temperatures, a rotation of the CO around the oxygen atom. The rotation implies that the C atom moves more than the O atom, explaining why it contributes more to the isotope effect. The barrier to binding depends on the distance from the oxygen fulcrum to the iron atom. Small variations in this distance, measured by the relevant Debye-Waller factor (Frauenfelder et al., 1979), may explain the distribution g(H) in barrier heights. The model explains some of the puzzles, but of course introduces new ones, such as "What is the role of the water molecule" in this elaborate dance.

Even with a detailed model as sketched here for the binding of CO to Mb, we lack predictive power. Future steps will involve use of the model to predict expected changes in the binding process on mutating one or more of the crucial residues lining the heme pocket and then performing the experiments to see how well the prediction works.

ACKNOWLEDGMENT

This contribution was written under the auspices of the U. S. Department of Energy.

REFERENCES

Alben, J. O., Beece, D., Bowne, S. F., Eisenstein, L., Frauenfelder, H., Good, D., Marden, M. C., Moh, P. P., Reinisch, L., Reynolds, A. H., and Yue, K. T. (1980) Phys. Rev. Lett. **44**, 1157–1160.
Alberding, N., Austin, R. H., Beeson, K. W., Chan, S. S., Eisenstein, L., Frauenfelder, H., and Nordlund, T. M. (1976) Science **192**, 1002–1004.
Ansari, A., Berendzen, J., Bowne, S. F., Frauenfelder, H., Iben, I. E. T., Sauke, T. B., Shyamsunder, E., and Young, R. D. (1985) Proc. Natl. Acad. Sci. USA **82**, 5000–5004.
Ansari, A., Jones, C. M., Henry, E. R., Hofrichter, J., and Eaton, W. A. (1994) Biochemistry **33**, 5128–5145.
Austin, R. H., Beeson, K. W., Eisenstein, L., Frauenfelder, H., and Gunsalus, I. C. (1975) Biochemistry **14**, 5355–5373.
Beece, D., Eisenstein, L., Frauenfelder, H., Good, D., Marden, M. C., Reinisch L., Reynolds, A. H., Sorensen, L. B., and Yue, K. T. (1980) Biochemistry **19**, 5147–5157.
Dickerson, R. E. and Geis, I. (1983) *Hemoglobin,* Benjamin/Cummings, Menlo Park, CA.
Doering, C. R. and Gadoua, J. C. (1992) Phys. Rev. Lett. **16**, 2318–2321.
Fleming, G. R. and Hänggi, P. Eds. (1993) *Activated Barrier Crossing.* World Scientific, Singapore.
Frauenfelder, H. and Wolynes, P. G. (1965) Science **229**, 337–345.
Frauenfelder, H., Sligar, S. S., and Wolynes, P. G. (1991) Science **254**, 1598–1603.
Frauenfelder, H., Bishop, A. R., Garcia, A., Perelson, A., Schuster, P., Sherrington, D., and Swart P. J. Eds. (1997) *Landscape Paradigms in Physics and Biology.* North-Holland, Amsterdam.
Frauenfelder, H., McMahon, B., Stojkovic, B. P., and Chu, K. (1998) in "Nonlinear Phenomena in Biological Systems." (In press.)
Frauenfelder, H., Petsko, G. A., and Tsernoglou, D. (1979) Nature **280**, 558–563.
Friedrich, J. (1995) Meth. Enzymol. **246**, 226–259.
Hagen, S. J., Hofrichter, J., and Eaton, W. A. (1996) J. Phys. Chem. **100**, 12008–12021.
Hänggi, P., Talkner, P., and Borkovec, M. (1990) Rev. Mod. Phys. **62**, 251–341.
Iben, I. E. T., Braunstein, D., Doster, W., Frauenfelder, H., Hong, M. K., Johnson, J. B., Luck, S., Ormos, P., Schulte, A., Steinbach P. J., Xie, A. H., and Young, R. D. (1989) Phys. Rev. Lett. **62**, 1916–1919.
Kleinert, T., Doster, W., Leyser, H., Petry, W. Schwarz, W., and Settles, M. (1998) Biochemistry, **37**, 717–733.
Kramers, H. A. (1940) Physica **7**, 284–304 (1940).
Landau, L. (1932) Phys. Z. Sovietunion **1**, 89.

Petsko G. A. and Ringe, D. (1984) Ann. Rev. Biophys. Bioeng. **13**, 331–371.

Quillin, M. L., Li, T., Olson, J. S., Phillips, G. N. Jr., Dou, Y., Ikeda-Saito, M., Regan, R., Carlson, M., Gibson, Q. H., Li, H., and Elber, R. (1995) J. Mol. Biol. **245**, 416–436.

Stein, D. L., Palmer, R. G., van Hemmen, J. L., and Doering, C. R. (1989) Phys. Lett. A **136**, 353.

Stryer, L. (1995) *Biochemistry.* W. H. Freeman, New York.

Stückelberg, E. G. C. (1932) Helv. Phys. Acta **5**, 369.

Zener, C. (1932) Proc. R. Soc. A. **137**, 696.

8

WHAT CAN NMR TELL US ABOUT PROTEIN MOTIONS?

Oleg Jardetzky

Stanford Magnetic Resonance Laboratory
Stanford University School of Medicine
Stanford, California 94305-5337

It is by now common knowledge that (1) motions, both random and concerted, are necessary for protein function and (2) high resolution NMR is the method of choice for studying motions in proteins at atomic resolution. In contrast, little attention has thus far been paid to the fundamental problem that the parameters measurable by NMR are not directly related to the mechanical parameters of functionally important motions, so that the relationship between NMR observables and the mechanics of the protein structure remains undefined. Most of the NMR literature on protein dynamics is devoted to the development and application of techniques to make the measurements more precise. This is, of course, desirable, but until the fundamental problem of relating the measurements to functionally important movements is solved, even the most sophisticated spectroscopic methods will produce little information that is relevant to biological function. It is the aim of this review to summarize the current state of our knowledge and the approaches being developed to solve this problem.

Functionally important motions in proteins that can be inferred from structural studies can be considered in three broad categories: (1) *en-bloc* translocations of ordered protein domains, which change the orientation of contact surfaces and are involved in allosteric control and signal transmission (Perutz, 1962; Monod, Wyman and Changeux, 1965; Lipscomb, 1994; Jardetzky 1996), (2) regional or segmental flexibility allowing induced fit (Koshland et al., 1966) and (3) ubiquitous thermal motion, which facilitates diffusion of ligands on protein surfaces and into the protein interior, permitting specific binding and catalytic function (Frauenfelder et al., 1979; Rasmussen et al., 1992). The common denominator of all are rotations and vibrations of individual bonds. The common denominator of all movements in the first category is selective rotation, and of those in the second category relatively free rotation around some of the bonds of the peptide backbone—in both cases changes of the Ramachandran angles phi and psi.

The time scales of processes reflected in the three classes are different. Random thermal motions are fast and occur on a pico to nano second time scale. The rates of *en-*

Protein Dynamics, Function, and Design, edited by Jardetzky *et al.*
Plenum Press, New York, 1998

bloc conformational transitions are slow and fall on a microsecond to second time scale. Segmental flexibility can be of different types and be observable on any time scale between and including the two. It is an open question whether the motions corresponding to the slow rates of conformational transitions are themselves slow, or whether they are as rapid as all thermal motion, but rare as events, yielding a slow kinetic rate.

The distinction between the kinetic rate—which is what is measured for processes on longer time scales–and the speed of the actual motion is fundamental, though not always observed–one frequently sees references to "slow motions" in proteins, where the actual measurement is a kinetic rate. We do not know whether there are any slow–low velocity–motions as such, but we cannot rule out their existence. We do know that there are slow motional events—an instantaneous transition (jump) between two long lived states is a motional event on the time scale defined by the lifetime of each of the states. The occurrence of motional events in proteins on time scales ranging over twelve orders of magnitude, from picoseconds to seconds has been established beyond all doubt (for a more detailed review see Jardetzky, 1996). Functionally the most important are the events on longer time scales.

NMR has gained preeminence among the methods for the study of protein dynamics, because it is the only experimental method by which motions in complex molecular structures, such as proteins, nucleic acids, lipid assemblies and polysaccharides can be monitored at atomic resolution. There are only two other methods that in principle permit the study of motion at atomic resolution - x-ray diffraction in real time and Molecular Dynamics (MD) simulations. Real time x-ray crystallography is not yet developed to the point at which it is producing many data relevant to protein dynamics—and it suffers from the severe limitation that the observations are made on crystals, where crystal forces are apt to affect dynamics much more than they affect structure. Molecular Dynamics is an indispensable tool for the visualization of motions in protein structures, but it is limited to timescales shorter than a few nanoseconds and its results are very dependent on the many assumptions made in each specific simulation. Its most severe limitation lies in the inaccuracy of the simplified potentials that have to be used to make simulations practical. All MD simulations require confirmation by experiment, since the method is not a source of experimental information about protein dynamics. All other experimental methods that have been used for the study of protein dynamics—many of which have an excellent time resolution for monitoring events on different time scales—have very poor, if any spatial resolution. Hence the unique role of NMR in this field.

Four types of NMR measurements allow inferences about protein dynamics:

1. Measurement of (mostly) heteronuclear (^{15}N-H, ^{13}C-H) relaxation parameters - T_1, $T_{1\rho}$, T_2, NOE.
2. Measurement of backbone proton exchange
   a. by H - D isotope substitution
   b. by relaxation methods
3. Chemical shift averaging and
4. Coupling constant averaging

The information provided by the first two is discussed below under separate headings. They are the main potential sources of data for a detailed description of the dynamics of larger proteins.

Chemical shift averaging has thus provided only qualitative information on dynamics at selected sites in the protein structure - such as methyl group rotation and the flipping of aromatic rings (Jardetzky, 1970; Campbell et al., 1976). One can infer no more than that the rate of averaging is faster than the chemical shift difference between two or more orientations, in frequency units (usually a few Hz and often unknown). It holds less promise for a detailed characterization of protein motions than either relaxation or proton exchange and is not discussed further. It is worth noting however that it provided some of the earliest incontrovertible—albeit qualitative—evidence for the occurrence of motional events in proteins on micro- to millisecond time scales.

Coupling constant averaging is of great interest in the study of small molecules, including small proteins, where coupling constants can be resolved. It reflects rotation around single bonds and often can be interpreted in terms of transitions between a small number of well defined states—e.g. the three staggered conformations in a Newman projection of a C-C bond. This type of analysis is of greatest interest in the study of ring puckering in carbohydrates and of amino acid side chain rotations. In larger proteins coupling constants are rarely resolved well enough to allow this type of analysis. Most proteins with interesting biological functions fall into this category and since the focus of this review is on the relationship between dynamics and biological function, this subject is also not further discussed. Modern approaches to the problem of defining rotamer populations from averaged coupling constants are described in recent papers by Dzakula et al. (1992; 1996).

Very detailed characterization of the dynamics of relatively small molecules has been achieved by a combination of relaxation and coupling constant measurements and energy minimization, as in the study of the small cyclic peptide antamanide by Richard Ernst and coworkers (Ernst, 1992 and references therein; Ernst and Ernst, 1994). A Boltzmann distribution of conformational states could be defined on that basis—containing no fewer than 1176 interconverting low energy structures. Rapid (<1.37 nsec) and slow (> 1.37 nsec) puckering of proline rings could be distinguished and a hydrogen bond with a lifetime of 25 microseconds has been found. It may not be practical to define such complete distributions of all conceivable structural variants as energy substates for larger proteins. At the same time, it may not be necessary for the understanding of functionally important conformational transitions. Most of the substates reflect small differences in side chain orientation, and will not be relevant to the larger structural and energetic differences between functionally different states of larger protein structures.

1. RELAXATION METHODS

Heteronuclear relaxation measurements are most commonly used for the study of protein dynamics because they are the easiest to interpret (Doddrell et al., 1972; Nelson et al., 1974). The relaxation of ^{15}N and ^{13}C nuclei is dominated by the covalently attached protons, with a smaller contribution arising (at higher fields) from their chemical shift anisotropy tensor. Such nearest neighbors at short fixed distances are rare in ^1H homonuclear relaxation and the analysis of the data is complicated by the multiple pathways allowed by the multiplicity of like neighbors (loosely called spin diffusion). Even for the "simple" two spin case the interpretation of the measured parameters in terms of motions is far from simple and this is evident from the fact that the literature on the subject is fraught with misconceptions, misunderstandings and hidden assumptions. It is therefore essential to preface any discussion with a definition of terms and of the conditions under which they apply.

The relaxation parameters for the observed nucleus are operationally defined by the Bloch equations, with the addition of the Overhauser term:
The Bloch Equations

$$\frac{dM_x}{dt} = \frac{M_x}{T_2}$$

$$\frac{dM_y}{dt} = -\gamma H_1 M_z - \frac{M_y}{T_2}$$

$$\frac{dM_z}{dt} = \gamma H_1 M_y - \frac{(M_z - M_0)}{T_1} \tag{1a}$$

The Nuclear Overhauser Effect

$$\text{in the steady state:} \quad -\rho_i \left(I_i - I_i^0 \right) + \sigma I_j^0 = 0 \quad \text{or} \quad I_i = I_i^0 + \left(\frac{\sigma}{\rho_i} \right) I_j^0$$

$$\text{and} \quad \text{NOE} \equiv \frac{I_i}{I_i^0} = 1 + \eta \quad \text{where} \quad \eta = \frac{\sigma I_j}{\rho I_i^0} \tag{1b}$$

The operational definition of the three parameters is pictorially represented in Fig. 1.

The simplest microscopic interpretation of the relaxation parameters has been given by Bloembergen (1948) for a pair of interacting nuclei attached to a rigid sphere randomly diffusing in solution. In this stochastic model—the nearest neighbor rigid rotor (RRNN) model—the internuclear vector describes a sphere whose origin can be taken at any point along the vector (usually at the observed nucleus). The random collisions responsible for the rotational diffusion occur within a range of frequencies and the diffusion can be described by a single exponential correlation function of the form:

$$G(\tau) = \overline{H_L(t)H_L(t+\tau)}$$

$$\text{if} \quad H_L(t) = H_L F(t)$$

$$\text{and} \quad H_L = \frac{\mu(3 \cos^2 \theta - 1)}{r^3}$$

$$\text{then} \quad G(\tau) = \langle H_L^2 \rangle \overline{F(t)F(t+\tau)} \tag{2}$$

where the coefficient tau—called the correlation time—can be taken as a measure of the average time the vector takes to rotate through one radian. The autocorrelation function itself describes the persistence of a stochastically diffusing vector in a given orientation.

The probability distribution of frequencies reflected in the motions of the vector as a result of random collisions is given by the spectral density function—a Fourier transform of the correlation function:

$$J(\omega) = \int_{-\infty}^{\infty} \overline{F(t)F(t+\tau)} \, e^{-i\omega t} dt \tag{3}$$

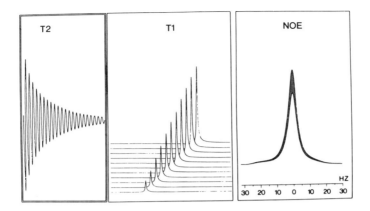

Figure 1. Pictorial representation of T_2, T_1 and NOE measurements.

Since relaxation represents the reorientation of the observed spin by the magnetic field of the other, if this field has components near their Larmor precession frequencies, the relaxation parameters can be expressed as functions of the spectral density function as follows:

$$\frac{1}{NT_1} = a^2\left[J(\omega_X - \omega_A) + 3J(\omega_A) + 6J(\omega_X + \omega_A)\right] + b^2 J(\omega_A)$$

$$\frac{1}{NT_2} = \frac{1}{2}a^2\left[4J(0) + J(\omega_X - \omega_A) + 3J(\omega_A) + 6J(\omega_X) + 6J(\omega_X + \omega_A)\right]$$

$$+ \frac{1}{6}b^2\left[4J(0) + 3J(\omega_A)\right]$$

$$NOE = 1 + \frac{\gamma_X}{\gamma_A}a^2\left[6J(\omega_X + \omega_A) - J(\omega_X - \omega_A)\right]NT_1$$

$$\text{Where} \quad a^2 = \frac{1}{10}\frac{\gamma_A^2\gamma_X^2\hbar^2}{\left(r_{AX}^3\right)^2} \quad \text{(for dipolar relaxation)}$$

$$\text{and} \quad b^2 = \frac{2}{15}\left[\gamma_A H_o\left(\sigma_\parallel - \sigma_\perp\right)\right] \quad \text{(for chemical shift anisotropy)} \qquad (4)$$

So far so good. The formulation for the RRNN model is valid and rigorous. Now, it is generally assumed that Eq. (4) are generally valid for all motions occurring within a complex structure such as a protein. This is equivalent to saying that all internal motions in proteins are stochastic in nature. On this assumption it is justified in saying that "the problem of calculating a relaxation rate is in essence the problem of determining the appropriate spectral density function" (Jardetzky, 1978). All attempts to account for internal motions in proteins to date have been based on this tacit assumption. However, if one remembers that a protein is an organized, structured, system, one can see that the assumption need not hold for a protein structure any more than it holds for other structured systems—such as a watch, or even a bridge, where specific motions and periodic or quasi-periodic resonant frequencies are built into the structure. In that case the J(ω) in Eq. (4) have to be replaced by a more general function

$$Y(\omega) = J(\omega) + L(\omega) \tag{5}$$

where L(ω) is a specific function appropriate for the description of the preprogrammed, as it were, coherent and correlated internal motions. The exact form of L(ω) depends on the nature of the concerted motions within the organized structure. Various approximate forms can be derived from specific models, such as the soliton model (Jardetzky and King, 1983), normal mode analysis (e.g. Brüschweiler and Case, 1996) or molecular dynamics (MD) simulations on extended time scales, such as MCDIS-MD (Monte Carlo in Dihedral angle Space-MD, Zhao and Jardetzky, 1994). The validity of specific models remains to be established. However, it is important to note that when we try to describe internal motion—especially motion on longer time scales—as a purely stochastic process, we are using Eq. (4) as a stochastic approximation, which is less than rigorous.

Of interest and in a class by itself is the procedure proposed by Peng and Wagner (1992a; 1992b; Lefèvre et al., 1996) for calculating all of the spectral density function terms in Eq. (4) from a sufficient number of measured relaxation parameters. This method of "spectral density mapping" is simply an experimental determination of the relative contributions of different spectral density terms to a given set of relaxation parameters. It does not attempt to relate these terms to any specific type of motion, nor is it inherently capable of doing so. The information content of the spectral densities calculated from experimental data is the same as that of the original data and both can be used to characterize relaxation behavior at different points in the structure without making any extraneous assumptions. It is in this lack of any added assumptions that spectral density mapping fundamentally differs from all of the procedures discussed below, all of which attempt to "explain" the spectral density function in terms of more or less clearly defined molecular motions. All are necessarily based on additional assumptions not derived from the data. It is therefore conceptually incorrect to regard spectral density mapping as an "alternative" (Nicholson et al., 1996) to any method which attempts to relate relaxation parameters to parameters of physical motions. Whether the relaxation behavior of two nuclei at two different points in the protein structure is best compared by comparing the spectral density terms or the raw data from which they have been derived is debatable. In neither case can anything be said about the nature of the underlying motions.

To relate the measured relaxation parameters—or the experimentally measured spectral density functions—to specific motions it is necessary to express the spectral density function in terms of parameters describing the motion. In principle the correct way of doing so is to begin with the correct and complete equations of motions for the system. In the case of a protein this would mean calculating the spectral density function from the trajectories of a Molecular Dynamics simulation—provided the simulation were based on a correct and complete set of the relevant equations of motion. This is where the problem lies. To make MD simulations computationally feasible one is forced to use simplified and approximate binary potentials to generate the forces driving the motions. This has two important consequences that limit the accuracy of the analysis—first the trajectories describe the motions only approximately, whereas the relaxation parameters reflect them exactly. Second - the omission of higher order interaction terms eliminates the potentially very important concerted motions from the simulation. We know that organized systems contain long range constraints and we can expect that proteins, as organized systems will undergo motions akin to lattice vibrations in a crystal. Although there have been attempts to account for more complex interactions in MD simulations, success in accounting for even the very fast motions responsible for NMR relaxation has been very limited (Zhao and Jardetzky, 1995). Not even a successful approximation for dealing with the slower motions has so far been achieved.

Approximate methods that have been developed to account for internal motions in proteins in terms of specific spectral density functions are summarized in Table I.

The alternative approaches that have been tried fall into two distinct categories: (1) specific models which attempt to describe the motion exactly (Hubbard, 1961; Woessner, 1962a; 1962b; Wittebort and Szabo, 1978; Howarth, 1979) and (2) model-free formulation which do not attempt a precise description of the motions, but try to capture some of their essential features (King and Jardetzky, 1978; Lipari and Szabo, 1982a; 1982b).

The fundamental difficulty in using specific models is that exact models have only been formulated for very simple motions, e.g., the rotation of a methyl group on a benzene ring. For more complicated motions, such as those in a protein, an exact model cannot easily be formulated as long as the nature of the underlying motions is not known—a catch 22 situation, since NMR is the main source of experimental information about these motions. An added difficulty stems from the fact that several simple models may account for a given set of relaxation data equally well, as had been shown in an early comparative study (King et al., 1978). Still, it appears that the old "wobble-on-a-cone" model (Woessner 1962b; King et al., 1978), most rigorously formulated by Howarth (1979), and extended to include anisotropic diffusion, provides an adequate approximate description for the motion of the NH vector in studies of protein backbone motions by ^{15}N relaxation (Zheng et al., 1995a; 1995b).

The fundamental fallacy of using model-free approaches is that they contain assumptions that are not applicable to protein structures and do not allow any real insight into the physical nature of the underlying motions. The original version (King and Jardetzky, 1978), which involves an expansion into as many motional terms as can be justified by the number of measurements, assumes that all motions are independent and stochastic, which may not be true for organized structures. The general method has found more application in the study of carbohydrates (Hricovini et al., 1992) than in the study of

Table I. Methods of determining
the spectral density function

(1) The rigid rotor nearest neighbor (RRNN) model
(2) The "effective correlation time" model (for known r)
(3) Specific models (e.g. wobble on a cone, for known r)
(4) Frequency analysis (for known r) (King and Jardetzky, 1978)

$$J(\omega) = \sum_i \frac{\alpha_i \overline{\lambda_i}}{\omega^2 + \lambda_i^2} \qquad \lambda = \tfrac{1}{\tau}$$

(5) Order parameter analysis (Lipari and Szabo, 1982a)

$$J(\omega) = \frac{2}{5}\left(\frac{S^2 \tau_M}{1 + (\tau_M \omega)^2} + \frac{(1 - S^2)\tau}{1 + (\tau\omega)^2} \right) \text{ with } \tau^{-1} = \tau_M^{-1} + \tau_e^{-1}$$

proteins, although it has been recently reintroduced into protein studies to account for observations for which it was difficult to account otherwise (Fujiwara and Nagayama, 1985; Ishima and Nagayama, 1995). The same assumption, as well as additional restrictive assumptions underlie the method most commonly used in the analysis of protein relaxation data—the model-free model of Lipari and Szabo (1982a; 1982b). The fundamental assumption of the method is that the net stochastic motion of an internuclear vector embedded in the protein structure is properly described by a correlation function which is a product of the correlation function for overall tumbling and the correlation function(s) for internal motion, i.e.,

$$\delta(\tau) = \delta^0(\tau)\delta_i(\tau) \tag{6}$$

For a single internal motion the spectral density function is then given by:

$$J(\omega) = \frac{2}{5}\left(\frac{S^2\tau_M}{1+(\tau_M\omega)^2} + \frac{(1-S^2)\tau}{1+(\tau\omega)^2}\right) \text{ with } \tau^{-1} = \tau_M^{-1} + \tau_e^{-1} \tag{7}$$

where S^2 is called an order parameter and is a nonspecific measure of the amplitude of motion. The great appeal of the model lies in its rigor and its simplicity. The problem is that it rigorously applies to a ball bearing - a rigid sphere with a single degree of internal motional freedom - but not to a protein which many degrees of internal motional freedom. Expansion of the model to two or more motions with two or more order parameters is possible and has been proposed (Clore et al., 1990), but then rigor is lost. As one surveys the literature dealing with the Lipari-Szabo model, it is clear that the analysis is not sensitive to the magnitude of the "internal" correlation time and that the only conclusions that can be drawn from the calculation of the order parameter are those that are fairly obvious on a careful examination of the original data: sometimes, in some parts of the peptide chain relaxation is less effective (small S^2) than in others (large S^2) presumably because of more extensive motion. The observed variations of the order parameter along the peptide chain in ^{15}N measurements generally tend to be small—a value typical for many proteins is 0.8 - and the magnitude of the order parameter does not tell us what the underlying motions are. It is worth noting however that the model-free model is mathematically equivalent to the "wobble-on-a cone" model as developed by Howarth, so that S^2 can be related to the half angle of the cone by

$$J(\omega) = A\frac{\tau_{iso}}{1+(\omega\tau_{iso})^2} + (1-A)\frac{\tau_p}{1+(\omega\tau_p)^2} \text{ where } A = \left[\frac{1}{2}\cos\alpha(1+\cos\alpha)\right]^2$$

and $\frac{1}{\tau_p} = \frac{1}{\tau_{iso}} + \frac{1}{\tau_e}$. The equation is identical to that of the model - free

model, with the order parameter : $S^2 = A$ \hfill (8)

as Lipari and Szabo had originally pointed out. In the Howarth interpretation an S^2 of 0.8. corresponds to the rather small half-angle of the cone of ~ 20°.

Here we encounter the most fundamental of all difficulties in the existing analysis of NMR relaxation data: Even if we knew how to describe the motion of a given internuclear vector exactly we still do not know how to relate it to any motion that may be involved in protein function. Before we deal with this fundamental issue we need to consider some common misconceptions that have hampered our thinking and some experimental data that may provide clues as to where to look for a better framework.

The most common misconception is that NMR relaxation parameters are only sensitive to fast motions— motions faster than the overall rate of diffusional tumbling. The origin of this misconception is Eq. (6), which says in words that the faster isotropic motion will average out all contributions of any slower motion. This is only partially true for exponential correlation functions (it would be rigorously true only if one were dealing with step functions). The overall correlation time does not provide a sharp cut off, and motions slower by as much as an order of magnitude than overall tumbling can still make a substantial contribution to the relaxation rate. More important is that Eq. (6) does not rigorously hold for proteins in the sense of accounting completely for the relaxation process. This has become quite apparent when it was found that a single spectral density function derived from the Lipari-Szabo formalism failed to account for all of the measured relaxation parameters simultaneously, which is required of any theory that is to be considered valid (Kay et al., 1989; Clore et al., 1990; Powers et al., 1992, among others). The discrepancy is usually removed by postulating an additional "exchange" term for T_2—indeed exchange between different chemical environments, conformational averaging, will contribute to line broadening—but other models, not starting from Eq. (6) are also possible. If the overall tumbling is anisotropic as it often is with larger structures, or if there is conformational heterogeneity, so that there is a variation in the orientation of the overall diffusion tensor within the structure with time, the relative contribution of slow motions can become even larger. As long as the exact nature of the motions in proteins remains unknown, one is not on safe ground arguing that motions excluded by a specific model cannot be detected or even cannot exist. In principle, it is possible for relaxation parameters to reflect motions on all time scales.

Another common misconception is that a precise quantitative analysis of a small set of relaxation data is meaningful. To be sure even a single relaxation measurement, T_1 or T_2, can provide a clue that a part of the protein is more mobile than another. However, the smaller the data set the more models can be fit to it. Any model will account for a single set of T_1 measurements at a single frequency. Internal consistency in accounting for an extensive data set—relaxation parameters on several nuclei at several frequencies—will be required to discriminate between them.

Most of the applications of simple models—including the model-free model of Lipari and Szabo—in fact most of the analyses of relaxation data published to date have involved small data sets. None of them have provided any insight into the nature of the underlying motions, beyond stating the obvious—hinges and chain terminals in protein structures tend to be more flexible than units of secondary structure. Most of the reported relaxation studies were carried out on proteins not known to undergo any functionally important structural changes, but also the simplicity of the theoretical analysis would preclude the identification of any motions that could be of functional interest. Relatively disordered regions have been detected by relaxation methods in the middle of several calcium binding proteins—calmodulin (Barbato et al., 1992), calbindin (Kördel et al., 1993; Akke et al., 1993), and troponin C (Kleerekoper et al., 1995). Without specifying the nature of the motions the relaxation findings support the "flexible hinge" model of the dumbbell-shaped molecules. A very extensive and careful relaxation study has been carried out on the 18 kD enzyme staphylococcal nuclease (Kay et al., 1989; Nicholson et al., 1992; 1996), using uniform ^{15}N labeling and selective ^{13}C labeling of alanines and leucines. It is apparent that the backbone of this protein is relatively constrained with order parameters of the order of $S^2 = 0.8$, except for the loop P42 - E57, where a slow motion was invoked to account for the data. Adding a ligand had little effect on the alanines, but selectively stabilized the side chains of Leucines 36, 37, 38, 89, 103, 108 and 125, near the binding site. Selective stabilization at binding sites has actually been known for a long time (Jardetzky, 1964), and is usually obvious from the raw data without requiring detailed analysis of the nature of the motion.

Of some interest are the findings in an ^{15}N relaxation study of the uniformly labeled, liganded and unliganded *trp*-repressor (Zheng et al. 1995b), an allosteric protein whose conformation is affected by the binding of the co-repressor L-tryptophan. A plot of the measured relaxation parameters T_1, T_2 and NOE as functions of the amino acid sequence is shown in Fig. 2.

It suggests that the molecule is relatively rigid throughout its structure. A more detailed analysis was carried out using the wobble-on-a-cone model of Howarth and the Lipari-Szabo formalism. The best fit to the data was obtained when the 1.7:1 anisotropy of

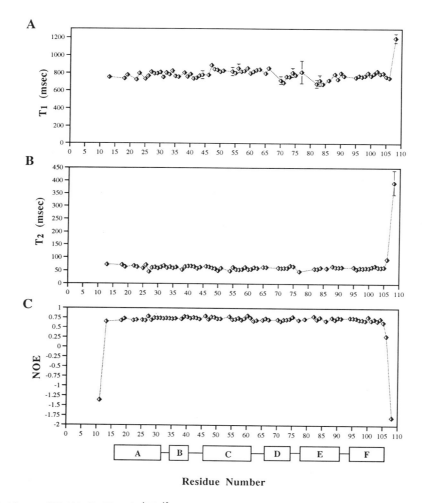

Residue Number

Figure 2. Measured T_1 (A), T_2 (B) and {^{1}H}-^{15}N NOE (C) values for the *trp* repressor plotted by residue number. The *trp* repressor secondary solution structure is indicated by the boxes at the bottom.

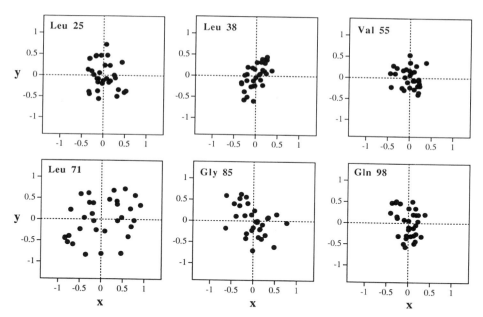

Figure 3. Cross sections through the plane perpendicular to the direction of the average NH vector for selected residues in *trp* repressor. The trajectories of NH vectors for each residue illustrate an internal motion pattern for each of the six helices. The NH vectors either form a cone (Leu 25, Val 55, Leu 71 and Gly 85) or a cone whose cross sections are ellipsoidal (Leu 38 and Gln 98).

the elongated molecule was taken into account in the Howarth model allowing for wobble in cones with a semiangles of $20.9 \pm 5.7.°$ The pattern of the wobble for several residues is shown in Fig. 3 as projections of the tip of the NH internuclear vector onto a plane perpendicular to its average orientation.

The calculated parameters are shown in Fig. 4.

The main conclusions are essentially the same as those that can be drawn from the raw data, with a slight hint of slightly greater flexibility in the hinge between the DNA binding helices D and E. An interesting finding was obtained on another DNA binding protein—the c-myb biding domain—by Ogata et al. (1995), where a flexibility detected by the need for an *ad hoc* "exchange" term in the relaxation analysis was correlated with the strength of binding to DNA. The more flexible wild type had a higher DNA binding constant than the more rigid V109L mutant.

On the whole, the results of NMR relaxation studies on proteins to-date have been relatively uninformative, as far as a detailed understanding of protein motions is concerned. The case of the *trp*-repressor was of special interest, because it is an allosteric protein, known to undergo functionally significant motions, and one might have expected that relaxation measurements would provide a clue as to the possible nature of the motions involved in the allosteric mechanism. However, it appears that relaxation measurements monitor for the most part ubiquitous thermal motion and not the motions involved in biological function. In cases such as the calcium binding proteins and the c-myb DNA binding protein the mere existence of flexibility or segmental disorder may facilitate induced fit and thus affect function. A more complete analysis of the motions—a better interpretation of relaxation parameters—will be needed to characterize the important motions in greater detail.

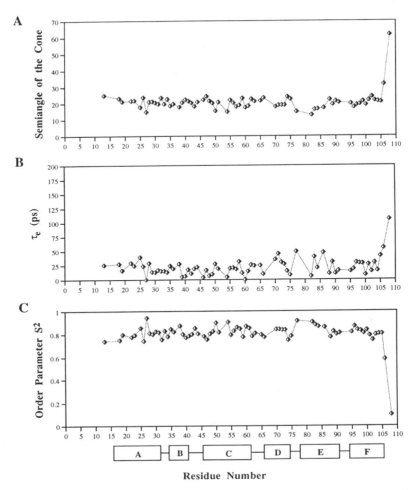

Figure 4. Plots of the semiangle of the cone (A), internal correlation time (B) and order parameter (C) as a function of residue number. The *trp* repressor secondary solution structure is indicated by the boxes at the bottom.

The direction in which the search for a better interpretation has to go is clearly set by the elementary requirement that the motion as described by an appropriate equation of motion and the motion detected by the nuclear relaxation process should be the same. Given that the complete and exact description is computationally intractable, the need is for approximations that capture the essence of the motions occurring in proteins better than the commonly used procedures. Whether procedures such as the recently developed MCDIS-MD (Monte Carlo in Dihedral Space with Molecular Dynamics, Zhao and Jardetzky, 1994; 1995) will provide the necessary answers remains to be established.

2. PROTON EXCHANGE STUDIES

A very important source of information on protein backbone dynamics are the measurements of NH proton exchange. Already the early studies of Linderstrøm-Lang (1955;

Hvidt and Nielssen, 1966) had established beyond all reasonable doubt that slow motions within the protein structure were responsible for the slow exchange of protein backbone protons with the surrounding solvent. Since it has become possible to assign NH signals in the NMR spectra to specific residues in the sequence, exchange can be monitored at individual sites along the polypeptide backbone. While in this case again, as in the case of relaxation parameters the relationship between the measured parameter and the underlying motions remains ill-defined, the wealth of data that is emerging will undoubtedly prove important in finding the appropriate relationship and in discriminating between alternative interpretations. Some progress in this direction is noticeable already.

As noted above, the exchange of backbone protons can be followed either by the simple isotope substitution method, if the exchange is on a time scale longer that 1-2 minutes, or by relaxation methods, if it is on a time scale of milliseconds to a few seconds.

In a simple isotope exchange experiment the protein is dissolved in D_2O and the intensity of NH peaks is followed by an HSQC experiment, as shown in Fig. 5.

The decay of peak intensity is as a rule a single exponential. The observed rates are generally orders of magnitude slower than the exchange rate from a free peptide - the "intrinsic" exchange rate for a given residue. Intrinsic exchange rates have been very extensively studied by Englander and coworkers (Englander and Mayne, 1992; Bai et al., 1993) and their pH and sequence dependence are well known. It is therefore possible to express the retardation of exchange by the protein structure by a "protection factor" P:

$$k_{obs} = P^{-1}k_{intrinsic} \qquad (9)$$

It is important to note that the definition of the protection factor as a ratio between two observables is independent of any assumptions concerning the exchange mechanism. The single exponential decay of line intensity is adequately described by an exchange from a single state and provides no clue as to the nature of the retardation. Differences in solvent accessibility would adequately account for the observed differences in protection factors for different backbone protons along the chain. However, the conventional interpretation of the retardation is in terms of the Linderstrøm-Lang (L-L) model, which assumes that the NH proton can exist in an inaccessible state, from which no exchange can occur, and an accessible state, from which exchange occurs at a rate equal to the intrinsic rate, i.e.,

$$NH_A \underset{k_{close}}{\overset{k_{open}}{\rightleftharpoons}} NH_B \underset{k'_{intrinsic}}{\overset{k_{intrinsic}}{\rightleftharpoons}} H_2O$$

2 limiting cases:

1) $k_{open} < 0.001 \text{ s}^{-1}$ $\quad k_{ex} \approx k_{open}$

2) $k_{close} \gg k_{open}$ $\quad k_{ex} \approx \dfrac{k_{open} k_{intrinsic}}{k_{close}}$ $\qquad (10)$

The original notion was that the inaccessible state corresponded to a proton hydrogen bonded in an alpha helix and the accessible state to a proton in an open helix (or beta sheet) - the terms "open" and "closed" states are still commonly used in discussions of the L-L model. Since, however, the analysis of experimental exchange results does not require the existence of an equilibrium, such as is postulated in the model, let alone a specific interpretation of the nature of the states in equilibrium, it is more meaningful to speak of a "protected" and "unprotected" state without further specifying their nature and properties. Much unnecessary confusion and controversy that has beset this field could have been avoided, if a clear distinction had been made from the beginning between the compelling inferences that can be drawn directly from the data without any specific assumptions and noncompelling inferences that can only be drawn on the assumption of a very specific model of interpretation. Failure to make this distinction has led to the formation of two rivaling models explaining the retardation of NH exchange in proteins: (1) the local unfolding model- following the original L-L concept (Englander and Kallenbach, 1984 and references therein), and (2) solvent penetration model (Rosenberg and Enberg, 1969; Woodward and Hilton, 1979). The simple isotope exchange experiment with a single exponential decay does not allow a distinction between these two models.

More detailed information on the exchange process can sometimes be obtained when exchange can be monitored by NMR relaxation methods. When the exchange rates are within an order of magnitude of the relaxation rate, as shown in Fig. 6, relaxation will be modified by exchange and the exchange rate can be estimated from the measured apparent relaxation parameters.

Magnetization recovery in a T_1 experiment in the presence of exchange is described by modified Bloch equations (McConnell equations):

$$\frac{dM_A}{dt} = -R_{1A}\left(M_A - M_A^{EQ}\right) - k_{open}M_A + k_{close}M_B$$

$$\frac{dM_B}{dt} = -R_{1B}\left(M_B - M_B^{EQ}\right) - k_{open}M_A - k_{close}M_B - k_{intr}M_B + k'_{intr}M_{H_2O}$$

$$\frac{dM_{H_2O}}{dt} = -R_{1H_2O}\left(M_{H_2O} - M_{H_2O}^{EQ}\right) - k'_{intr}M_{H_2O} + k_{intr}M_B \tag{11}$$

whose general solution is bi-exponential

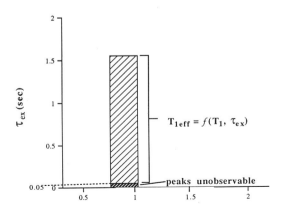

Figure 6. τ_{ex} window in an NH T_1 experiment.

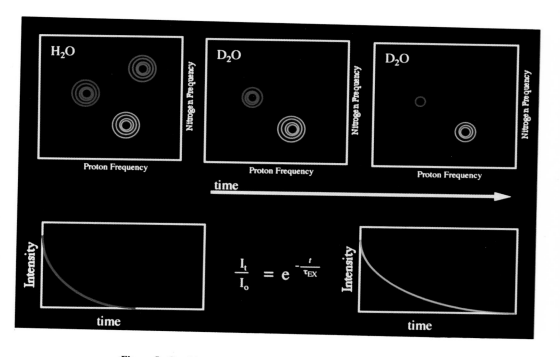

Figure 5. Graphic representation of deuterium exchange experiments.

$$M_T(t) = M_A(t) + M_{B0}$$

$$M_A(t) = M_{A0}(1 - P_A) + C_1 \exp(-\lambda_1 t) + C_2 \exp(-\lambda_2 t)$$

$$M_B(t) = M_{B0}(1 - P_B) + D_1 \exp(-\lambda_1 t) + D_2 \exp(-\lambda_2 t) \tag{12}$$

Indeed, bi-exponential recovery is frequently observed in such experiments, as shown in Fig. 7.

It is important to note that unlike the case of a single exponential decay, which does not require that more than one state be invoked in accounting for the data, bi-exponential recovery implies the existence of at least two states. The general solution of the kinetic equations describing the Linderstrøm-Lang model is bi-exponential—a fact that is often

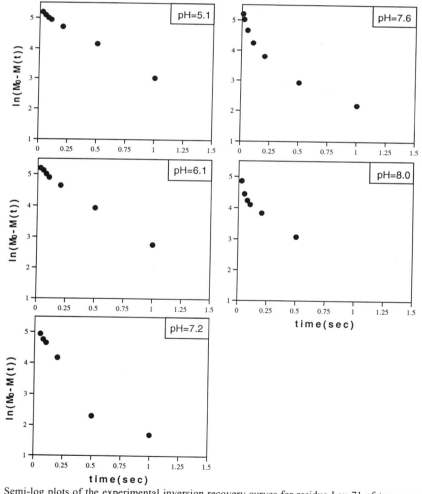

Figure 7. Semi-log plots of the experimental inversion recovery curves for residue Leu 71 of *trp* repressor. The experiments were taken with no solvent saturation and measured at several pH values, as indicated in the figure.

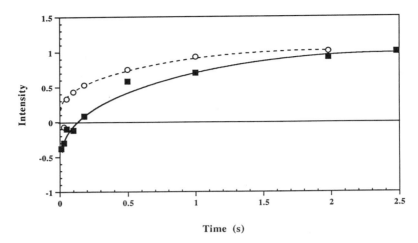

Figure 8. Magnetization recovery rates of for exchanging protons with (■) and without (○) presaturation.

forgotten, since the model is usually discussed in the simplified form that is sufficient for the description of single exponential exchange data.

The relaxation experiment to measure exchange rates can be carried out without or with simultaneous solvent saturation (normally employed to suppress the water signal, Krishna et al., 1979; Spera et al., 1991). The rates of magnetization recovery for the exchanging proton will be different in the two cases, allowing an additional independent estimate of the exchange rate. This is shown in Fig. 8.

Eq. (12) represents the general case for the analysis of NMR exchange data in terms of the Linderstrøm-Lang model. Important is that in cases in which bi-exponential exchange is observed it allows a direct calculation of all kinetic parameters from experimental data. It also permits a systematic classification of cases in which simplifying conditions

Table II. pH dependencies of amide proton relaxation recovery profiles for experiments without solvent saturation

		Single exponential	Double exponential
pH independent	(a)[1]	$k_{open} < 0.01\ k_{close}$, $k_{open} < 0.01\ R_1$	N/A[3]
	(b)[2]	$k_{open} < 0.01\ R_1$	
pH dependent	(a)[4]	$k_{open}, k_{close} > 100\ R_1$	(a)[6] $0.1 < \dfrac{k_{open}}{k_{close}} < 10$, $k_{open}, k_{close} < 10\ R_1$
	(b)[5]	$k_{open} > 100\ k_{close}$	(b)[7] $0.1 < \dfrac{k_{open}}{k_{close}} < 10$, $0.1\ R_1 < k_{open}, k_{close} < 10\ R_1$

[1]The closed population is favored.
[2]Only the closed state amide proton is observed.
[3]This situation will not occur if the amide proton dipolar relaxation can be approximated as a single-exponential recovery process.
[4]It is irrelevant which states are observed.
[5]The open state is favored.
[6]Sum of the open and the closed states is observed as a single peak. The bi-exponential recovery curve is characterized with a strong pH dependent fast initial rate followed by a weakly pH-dependent or pH-independent slow rate.
[7]Closed state is observed alone. The recovery profile is weakly pH dependent, mostly single-exponential. The recovery curve becomes bi-exponential only when the k_{open}, k_{close} are comparable to the k_{int}. The bi-exponential recovery curve is characterized by a slow-fast recovery pattern.

apply and of alternative interpretations which can apply to a given experimental finding, especially in the study of the pH dependence of exchange. These are shown in Fig. 9 and Table II respectively.

A more detailed discussion of specific cases and the derivation of the general case can be found in the papers by Zheng et al., 1995b and Gryk et al., 1995.

An application of the foregoing type of analysis to the backbone proton exchange in the *trp*-repressor has led to several important conclusions (Finucane and Jardetzky, 1995; Gryk et al., 1995). In this protein single exponential exchange on a time scale of minutes to hours is observed for the core of the molecule—helices ABCF—while bi-exponential exchange on a millisecond to second time scale is seen in the DNA-binding helix-turn helix of helices D and E. This finding by itself indicates that the DNA binding region is relatively unstable, or flexible, and this instability or flexibility can facilitate adjustment to the DNA ligand i.e., induced fit. Direct calculation of the "intrinsic" exchange rate i.e., the exchange rate from the "unprotected" state, from experimental data led to the startling result

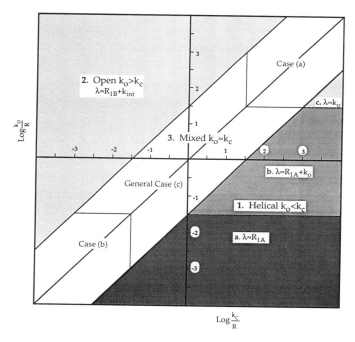

Figure 9. The parameter space outlining possible cases of the generalized equations, Eqs. (7) to (11). R_{1A} has been assumed to be approximately equal to R_{1B} for simplicity. (1) Predominantly helical. $k_{close} \gg k_{open}$. As only one state is being observed, relaxation will be single exponential. (1a) $k_{open} < R_1$ and $\lambda \approx R_{1A}$, (1b) $k_{open} > R_1$ and $\lambda \approx R_{1A}$ $+ k_{open} k_{eff} /(k_{close} + k_{eff})$. ($k_{eff} = k_{int} + R_{1B}$). Solvent exchange may occur at extremely high pH. The recovery will, however, remain single exponential. (2) Predominantly open. $k_{open} \gg k_{close}$. Again, as only one state is being observed, relaxation will be single exponential ($\lambda \approx k_{eff}$). pH dependence is to be expected. (3) $k_{open} \approx k_{close}$. Neither state predominates, and biexponentiality is possible as the open and closed states have comparable populations. (3a) "High motility" limit. $k_{open}, k_{close} \gg k_{int}$. The open and the closed states rapidly interconvert. A single, pH-dependent exponential will result. The exact extent of this region depends strongly on the pH. (3b) "Low motility" limit. The opening and closing rates are extremely slow and the two states relax independently of each other. When $k_{int} < R_1$, single exponential recovery is expected ($R_{1A} \approx R_{1B}$). When $k_{int} > R_1$, biexponential recovery will occur. (3c) The most general case. All rates are approximately equal and no simplifications hold. NMR relaxation methods are the methods of choice for the determination of the rates in this region.

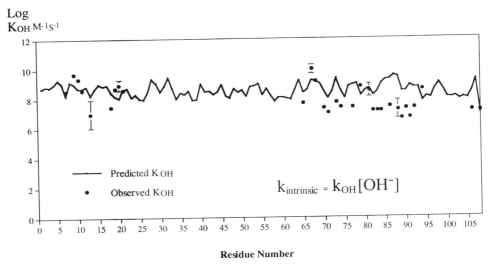

Log
K_{OH} M⁻¹s⁻¹

Residue Number

Figure 10. Comparison of calculated (——) and observed (●) $k_{intrinsic}$ ($k_{intrinsic}$ = k_{OH} [OH⁻]) values for *trp* repressor. The calculated values were obtained using the method of Bai et al. (1993).

that the experimentally determined "intrinsic" rate was slower by as much as two orders of magnitude than the "intrinsic" rate as determined from data on free peptides by the method of Bai et al. (1993). This is shown in Fig. 10.

There are two possible interpretations of this finding: (1) the "unprotected" state is not completely unprotected and the usual assumption made in the L-L model that exchange from this state occurs at the same rate as in free peptides does not apply, or (2) the simple L-L model does not apply and has to be extended to postulate two successive equilibria. The exchange data by themselves do not allow a distinction between these two interpretations. We favor the first, because there is evidence from structural studies that despite their instability, the DNA binding helices retain their average helical conformation most of the time (Zhang et al., 1994; Gryk et al., 1995) i.e., they do not unfold and refold as a whole, as has been suggested (Spolar and Record, 1994). At the same time there is no other evidence for the existence of additional states involved in the more complex equilibria. The finding suggests that when the concept of a protection factor is applied to the L-L model, one can consider protection in each of the states—the "inaccessible" and the "accessible" separately. Thus we can define the protection factors as follows:

$$P^{-1} = \frac{k_1}{(k_2 + k_3)} \quad \text{or for the EX}_2 \text{ mechanism:} \quad P^{-1} = \frac{k_1}{k_2} \quad \text{assuming } k_3 = k_{intrinsic} \tag{13}$$

when this assumption does not hold, we would have:

$$k_3 = P_b'^{-1} k_{intrinsic} \tag{14}$$

and for the overall protection factor:

$$P' = P_a' P_b' \text{ if } k_3 \neq k_{intrinsic} \tag{15}$$

The two protection factors can be measured only in the case of biexponential exchange, but they are useful reminders that the conventional assumption when applying the L-L model to single exponential exchange, i.e. that $k_3 = k_{intrinsic}$, need not be generally correct.

A second intriguing finding apparent in Fig. 10 is that the local protection factors - the ratios of $k_{intrinsic}$ and k_3 - are not identical along the chain. This is most important as it indicates that neither helix unfolds as a unit, but that the protons inside them become accessible to the solvent at individual and different rates. If we assume that accessibility results from the breakage of hydrogen bonds, comparison of local protection factors for adjacent NHs along the chain allows us to distinguish between different mechanisms of "unfolding," as shown in Fig. 11.

Examination of protection factors for the stable core helices ABCF, where backbone proton exchange is slow, also show that they differ by more than an order of magnitude (Table III).

This finding is incompatible with the concept originally derived from global monitoring of exchange i.e., that the helices unravel as a unit in order to permit exchange.

The study of proton exchange at individually identified sites within the structure, which is possible by NMR, allows us to distinguish between the different mechanisms of exchange in proteins that have been proposed. The "local unfolding" mechanism, using a simplified formulation of the L-L model requires that there be a rate limiting unfolding step, with concerted breakage of the secondary structure. The "solvent penetration" mechanism attributes the differences in rates to differences in solvent access. The distinction between the proposed mechanisms of backbone proton exchange is indicated in Fig. 12. The cited findings eliminate at least the most naive interpretation of the "local unfolding" model for proton exchange as a possible mechanism. If there is "local unfolding," it must be limited to at most one or two hydrogen bonds at a time. In the case of the *trp*-repressor not only the inequality of protection factors along the chain, but also the fact that a large fraction of the backbone protons are near

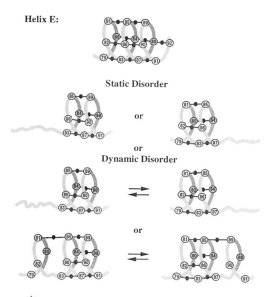

Figure 11. A few dynamic alternatives for "50% disorder" detected by static methods.

Table III. Comparison of the base-catalyzed exchange rates of amide protons in *trp*-repressor between the apo- and holo-forms

Residue	pH	$\log k_{OH}$ (log M^{-1} s^{-1})			Protection factor[1]		
		Apo	Holo	Calculated	Apo	Holo	Holo/apo
A9	5.20	9.54 (9.16–9.74)	9.82 (9.80–9.83)	9.02	0.3	0.2	0.5
A10	5.20	9.42 (9.33–9.49)	9.36 (9.34–9.37)	8.72	0.2	0.2	1.1
N40	7.60	7.12 (7.01–7.20)		9.00	75.9		
D46	7.60	7.88 (7.69–8.01)		8.18	4.2		
S67	6.05	8.77 (8.72–8.82)	8.85 (8.85–8.85)	9.20	2.7	2.2	0.8
E70	6.05	8.28 (8.25–8.31)	8.38 (8.04–8.57)	8.43	1.4	1.1	0.8
L71	7.60	7.56 (7.51–7.61)	7.24 (7.11–7.34)	7.99	2.7	5.6	2.1
L75	7.60	7.56 (7.50–7.62)	7.14 (7.07–7.20)	7.99	1.3	7.1	5.5
G76	6.05	8.75 (8.71–8.78)	8.79 (8.73–8.84)	8.78	1.1	1.0	0.9
A77	6.05	8.37 (8.33–8.40)	<7.64 (. -7.94)	8.89	3.3	>17.8	>5.4
G78	6.05	8.86 (8.83–8.88)	8.86 (8.83–8.88)	8.99	1.3	1.3	1.0
T81	6.05	8.87 (8.84–8.90)	8.66 (8.56–8.74)	8.65	0.6	1.0	1.6
I82	7.60	7.60 (7.56–7.64)	7.28 (7.04–7.44)	8.19	3.9	8.1	2.1
T83	7.60	7.77 (7.70–7.84)	7.54 (7.29–7.69)	8.42	4.5	7.6	1.7
G85	7.60	7.49 (7.44–7.53)	7.24 (7.20–7.27)	9.21	52.5	93.3	1.8
L89	7.60	7.74 (7.67–7.81)	7.07 (7.02–7.11)	8.44	5.0	23.4	4.7
A91	6.05	8.43 (8.14–8.60)	7.93 (7.87–7.98)	8.84	2.6	8.1	3.2
A92	7.60	6.80 (6.74–6.86)	6.51 (6.38–6.61)	8.72	83.2	162.2	1.9
V94	7.60	7.62 (7.56–7.67)	7.90 (7.87–7.93)	7.78	1.5	0.8	0.5
E95	7.60	7.55 (7.40–7.65)	7.44 (7.27–7.57)	8.07	3.3	4.3	1.3
D108	7.60	7.20 (7.17–7.23)	7.01 (7.01–7.02)	6.92	0.5	0.8	1.5

[1] The protection factor is the amount by which the base-catalysed exchange rate is reduced compared with the value calculated for small peptides. P=k_{OH} calculated/k_{OH} measured. The base-catalyzed exchange rates have been measured by fitting the fast phase of the biexponential at a single pH (listed in the Table) as described in the text. Because of the limited data set used (3 pH's) and because of possible non-linearity of the exchange rates with pH, the data at all 3 pHs were not fitted simultaneously to a model which assumes this to be the case (Gryk *et al.*, 1995). To ensure that the results remain comparable to the holorepressor, the data were for the holorepressor was refitted in the same manner, *i.e.*, at only one pH, that chosen which gave the best measurable rate for the aporepressor. The data were also fitted with a calculated pulse efficiency from the Jump-Return excitation profile. In most cases the differences in the calculated values for the holorepressor are small.

1. Solvent Penetration Model

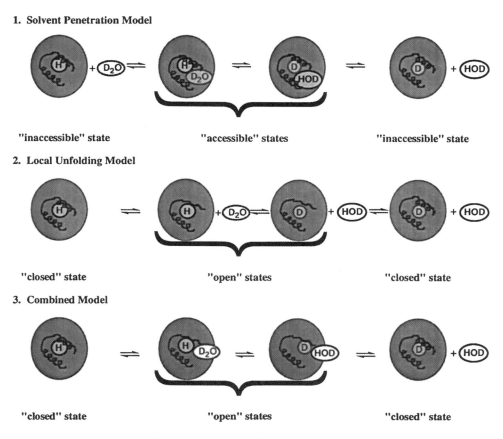

"inaccessible" state   "accessible" states   "inaccessible" state

2. Local Unfolding Model

"closed" state   "open" states   "closed" state

3. Combined Model

"closed" state   "open" states   "closed" state

Figure 12. Possible models for HD backbone proton exchange.

the surface suggest that solvent access and local dynamics jointly play a decisive role in determining the exchange rate. The protons near the surface can exchange by a penetration/diffusion mechanism, without involving a rate limiting unfolding step.

Evidence for a local unfolding of the tertiary, rather than secondary, structure as a contributing factor in accelerating the exchange rate of some residues at low pH has nevertheless been found in studies of the pH dependence of proton exchange. Normally, the exchange rate increases monotonically with pH in the range pH 5–9, as shown in Fig. 13A and B. For some residues, however, it goes through a minimum and increases as the pH is lowered from 6.2 to 5.5, as seen in Fig. 13C and D. This increase is accompanied by significant changes in the chemical shift in the surroundings of the affected NHs, strongly suggesting a partial disintegration of the tertiary structure, as illustrated in Fig. 14.

One of the most significant observations that has now been made on several proteins is the stabilization of the structure on ligand binding. For the *trp*-repressor case this is evident from the increase in protection factors throughout the structure upon binding of the co-repressor tryptophan (column 3 of Table III). The damping of exchange at sites distant from the binding sites is especially interesting, as it is not accompanied by a significant change in the intervening structure. As a consequence, this action at a distance must involve a modulation of the dynamics, but its exact nature remains to be elucidated.

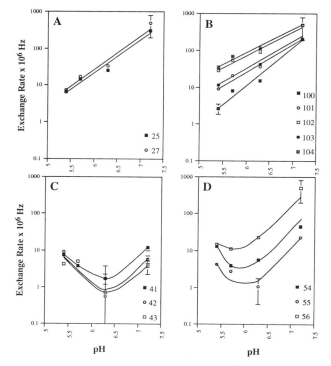

Figure 13. Dependence of the exchange rate (x 10^6 s^{-1}) on pH for selected residues in *trp* repressor helix A (A), helix F (B), the B-C turn (C) and helix C (D).

Of far reaching significance is also the finding that a difference in the dynamics between the wild type *trp*-repressor and a point mutant is associated with a difference in functional properties (Gryk and Jardetzky, 1996). Substitution of a valine for a glycine in the hinge of the DE helix-turn-helix greatly reduces the backbone proton exchange rates in the DNA binding domain, as seen in Fig. 15.

At the same time, the hinge mutant AV77 becomes a superrepressor for the *trp*-operon. This change in function is known not to be associated with a change in the binding constant to the *trp* operator DNA (Hurlburt and Yanofsky, 1990), but to have a complex kinetic mechanism (Arvidson et al., 1993). It results from a change in the specificity—or selectivity—for different DNA sequences. The mutant is more effective as a repressor for some operators, but less effective for others, as seem in Table IV.

All of the cited findings for the present allow only a qualitative interpretation. Nevertheless it is fair to say that the study of backbone proton exchange has so far produced more interesting observations bearing on protein dynamics and function than the study of relaxation phenomena. This is not too surprising, since most biologically important protein functions occur on time scales that are comparable to the time scale of proton exchange, rather than to the time scale of thermal motions. There is no doubt that proton exchange reflects changes of states—and in that sense motions—that occur on the same time scale as the exchange itself. However, the conceptual framework for defining these states and relating the two remains to be constructed.

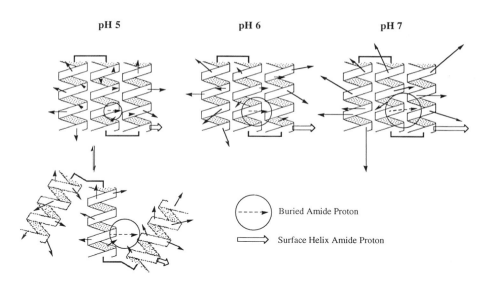

Figure 14. Two-process model for exchange from *trp* repressor. At pH 5 the repressor partially unfolds, increasing the exchange rates of the most slowly exchanging protons. At this pH, for the most slowly exchanging protons, $k_{obs} \approx k_{unfolding}$. At pH 7 the protein is stable and does not significantly unfold, and the exchange rates depend on the degree of burial of the amide protons from solvent. The arrows represent the exchange of amide protons, with the length of the arrows approximately representing the logarithm of the exchange rates. The dashed arrow represents a typical surface-helix amide proton, and shows the approximately linear increase in $\log(k_{obs})$ with pH. The large arrow represents a typical buried amide proton exchanging, and first decreases in k_{obs}, before increasing, with respect to pH.

3. SUMMARY AND CONCLUSIONS

NMR studies of protein dynamics have made an important contribution in providing direct experimental evidence for the existence of a variety of motions in proteins. Thus far this evidence can only be considered as having only qualitative significance, even when couched in quantitative terms. The basic findings are (1) some segments of the polypeptide backbone in a protein are more mobile, flexible or fragile, while most of the backbone is relatively rigid, (2) proteins can differ widely in the relative proportions of rigid and flexible regions in their structures and (3) side chain rotations are ubiquitous, though more constrained on the inside of globular proteins than on their surface. Saying that a segment has a large (or small) spectral density term or a large (or small) order parameter is not saying much more than that. To make a meaningful quantitative analysis one would need to know the nature and characteristics of the underlying motions. If one relies on models, such as the wobble-on-a-cone model, one can account for the anisotropy of the overall motion and semi-quantitatively confine the stochastic jumps of the observed internuclear vector to a cone of defined semi-angle. This still falls short of saying anything about the underlying motions which cause the jumps. Similarly, the opening of hydrogen bonds that is (by definition) necessary to account for backbone proton exchange from intact secondary structures, can for the present only be described qualitatively, and provides no insight into the motions responsible for it.

The missing relationships between the observables and the underlying motions of the protein structure are bound to emerge, in the case of both the relaxation parameters

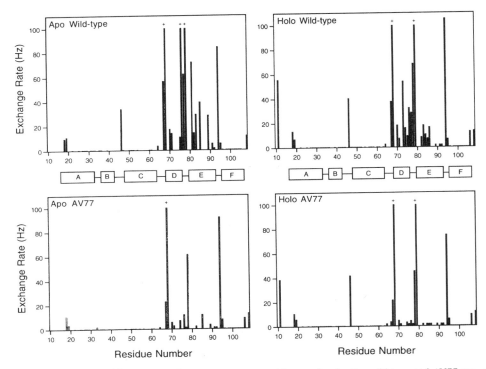

Figure 15. Measured amide proton exchange rates versus residue number for the wild-type and AV77 apo- and holorepressors. Exchange rates were measured at 45°C at pH 7.6.

and the backbone exchange rates, from a more complete and correct analysis in terms of molecular dynamics. The reported poor agreement between measured NMR relaxation parameters and those calculated from relevant MD trajectories indicates only that the conventional formulation of the theoretical parameters in the simulations was not adequate to the task. A closer examination of the correlations apparent in the experimental data is

Table IV. Repression of three promoter/operator-*lacZ* translational fusions by wild-type repressor and superrepressor AV77 *in vivo*

	β-galactosidase activity in strains with fusion[a]					
	trpL'-'lacZ[b]		*aroH'-'lacZ*[c]		*trpR'-'lacZ*[d]	
Plasmid[c]\media	−Trp	+Trp	−Trp	+Trp	−Trp	+Trp
None[f]	12,400	12,000	190±15	190±52	210±33	210±66
pRLK13[f](wild-type)	500±80	5±1	70±5	25±4	65±4	5±0
pAV77	130±11	4±0	115±8	25±5	75±6	3±0

[a] β-galactosidase assays were performed essentially as described by Miller (1972).
[b] λTLF1, a λ derivative containing a *trp* p/o *trpL'-'lacZ* fusion
[c] λRLK22, a λ derivative containing an *aroH* p/o *aroH'-'lacZ* fusion
[d] λTRL3, a λ derivative containing a *trpR* p/o *trpR'-'lacZ* fusion
[e] Each of the listed plasmids was present in W3110 Δ*lacU169 tnaA2 trpR2* strains lysogenic for each λ derivative.
f The 6 values on each of the first 2 lines were reported in Klig et al. (1988). The values for pAV77 were obtained at the same time, but were not published. The values presented are the averages from at least four separate experiments.

bound to lead to better approximations, even if the exact solution is beyond reach. There is no doubt that the approach is correct *in principle,* i.e. that *in principle* it must be possible to calculate the measured NMR parameters from the correctly formulated equations of motion of a protein structure. Once an even approximately complete and correct formulation is found, it will be the NMR data that will allow us to understand the nature of internal motions in proteins and their relation to function.

ACKNOWLEDGMENTS

This research was supported by NIH Grant GM33385.

REFERENCES

Akke, M., Skelton, N. J., Kördel, H., Palmer, A. G. III and Chazin, W. J. (1993). Biochemistry **32**, 9832–9844.
Arvidson, D. N., Pfau, J. , Hatt, J. K., Shapiro, M., Pecoraro, F. S. and Youderian, P. (1993). J. Biol. Chem. **268**, 4362–4369.
Bai, Y., Milne, J. S., Mayne, L., and Englander, S. W. (1993). Proteins: Structure, Function, and Genetics **17**, 75–86.
Barbato, G., Ikura, M., Kay, L. E., Pastor, R. W. and Bax, A. (1992). Biochemistry **31**, 5269–5278.
Bloembergen, N. (1948) *Nuclear Magnetic Relaxation.* Ph.D. Thesis, Leiden, Schotanus & Jens N.V., Utrecht, The Netherlands.
Brüschweiler, R. and Case, D. A. (1994) Phys. Rev. Lett **72**, 940–943.
Campbell, I. D., Dobson, C. M., Moore, G. R., Perkins, S. J. and Williams, R. J. P. (1976). FEBS Lett. **70**, 96–101.
Clore, G. M., Driscoll, P. C., Wingfield, P. T. and Gronenborn, A. M. (1990). Biochemistry **29**, 7387–7401.
Doddrell, D., Glushko, V. and Allerhand, A. (1972). J. Chem. Phys. **56**, 3683.
Dzakula, Z., Westler, W. M., Edison, A. S. and Markley, J. L. (1992). J. Am. Chem. Soc. **114**, 6195–6199.
Dzakula, Z., Westler, W. M. and Markley, J. L. (1996). J. Magn. Reson. Ser. B **111**, 109–126.
Englander, S. W. and Kellenbach, N. R. (1984). Quart. Rev. Biophys. **16**, 521–655.
Englander, S. W. and Mayne, L. (1992). Annu. Rev. Biophyps. Biomol. Struct. **21**, 243–265.
Ernst, M. and Ernst, R. R. (1994). J. Magn. Res., Ser. A **110**, 202–213.
Ernst, R. R. (1992). Angew. Chem. Int. Ed. Engl. **31**, 805–823.
Finucane, M. D. and Jardetzky, O. (1995). J. Mol. Biol. **253**, 576–589.
Finucane, M. D. and Jardetzky, O. (1996). Prot. Sci. **5**, 653–662.
Frauenfelder, H., Petsko, G. A. and Tsernoglou, D. (1979). Nature **280**, 558–563.
Fujiwara, T. and Nagayama, K. (1985). J. Chem. Phys. **83**, 3110–3117.
Gryk, M. R. and Jardetzky, O. (1996). J. Mol. Biol. **255**, 204–214.
Gryk, M. R., Finucane, M. D., Zheng, Z. and Jardetzky, O. (1995). J. Mol. Biol. **246**, 618–627.
Gryk. M. R., Jardetzky, O., Klig, L. S. and Yanofsky, C. (1996). Prot. Sci. **5**, 1195–1197.
Howarth, O. W. (1979). J. Chem. Soc. Faraday Trans. II **75**, 863–873
Hricovini, M., Shah, R. N. and Carver, J. P. (1992). Biochemistry **31**, 10018–10023.
Hubbard, P. S. (1961). Rev. Mod. Phys. 33, 249.
Hurlburt, B.K., and Yanofsky, C. (1990). J. Biol. Chem. **265**, 7853–7858.
Hvidt, A. and Nielsen, S. O. (1966). Adv. Prot. Chem. **21**, 287–386.
Ishima, R. and Nagayama, K. (1995). *Biochemistry* **34**, 3162–3171.
Jardetzky, O. (1996). Prog. Biophys. Mol. Biol. **65:3**, 171–219.
Jardetzky, O. (1964). Science **146**, 552–553.
Jardetzky, O. (1970). In: *Molecular properties of drug receptors* (Ciba Found. Symp.), (R. Porter and M. O'Connor, eds.), J. and A. Churchill, London, pp. 113–132.
Jardetzky, O. (1978). Life Sciences **22**, 1245–1252.
Jardetzky, O. and King, R. (1983) In: *Mobility & Function in Proteins & Nucleic Acids* (Ciba Foundation Symposium *93*), Pitman, London, 291–309.
Kay, L. W., Torchia, D. A. and Bax, A. (1989). Biochemistry **28**, 8972–8979.
King, R. and Jardetzky, O. (1978). Chem. Phys. Lett. **55**, 15–18.
King, R., Maas, R., Gassner, M., Nanda, R.K., Conover, W.W. and Jardetzky, O. (1978). Biophys. J. **24**, 103–117.

Kleerekoper, Q., Howarth, J. W., Guo, X., Solaro, R. J. and Rosevear, P. R. (1995). Biochemistry **34**, 13343–13352.

Klig, L. S., Carey, J. and Yanofsky, C. (1988). J. Mol. Biol. **202**, 769–777.

Kördel, J., Skelton, N. J., Akke, M. and Chazin, W. J. (1993). J. Mol. Biol. **231**, 711–734

Koshland, D. E., Nemethy, G. and Filmer, D. (1966). Biochemistry **5**, 365–385.

Krishna, N.R., Huang, D.H., Glickson, J.D., Rowan III, R., and Walter, R. (1979). Biophys. J. **26**, 345–366.

Lefèvre, J.-F., Dayie, K. T., Peng, J. W. and Wagner, G. (1996). Biochemistry **35**, 2674–2686.

Linderstrøm-Lang, K. (1955). Chem. Soc. Spec. Publ. **2**, 1–20.

Lipari, G. and Szabo, A. (1982a). J. Am. Chem. Soc. **104**, 4546–4559.

Lipari, G. and Szabo, A. (1982b). J. Am. Chem. Soc. **104**, 4559–4570.

Lipscomb, W. N. (1994). Adv. Enzymol. **68**, 67–152.

Miller, J. (1972) *Experiments in molecular genetics*. Cold Spring Harbor Laboratory Press, Cold Spring Harbor, NY.

Monod, J., Wyman, J. and Changeux, J. P. (1965). J. Mol. Biol. **12**, 88–118.

Nelson, D. J., Cozzone, P. J. and Jardetzky, O. (1974). In: *Molecular and Quantum Pharmacology*, (E. Bergmann and B. Pullman, eds.), D. Reidel Publ. Col., Dordrecht-Holland, pp. 501–513.

Nicholson, L. K., Kay, L. E., Baldisseri, D. M., Arango, J., Young, P. E., Bax, A. and Torchia, D. A. (1992). Biochemistry **31**, 5253–5263.

Nicholson, L.K., Kay, L.E. and Torchia, D. A. (1996). In: *NMR Spectroscopy and its Application to Biomedical Research*, (S. K. Sarkar, ed.), Elsevier, Amsterdam, The Netherlands, pp. 241–279.

Ogata, K., Morikawa, S., Nakamura, H., Hojo, H., Yoshimura, S., Zhang, R., Aimoto, S., Ametani, Y., Hirata, Z., Sarai, A., Ishii, S. and Nishimura, Y. (1995). Nature Struct. Biol. **2**, 309–320.

Peng, J. W. and Wagner, G. (1992a). J. Magn. Reson. **98**, 308–332.

Peng, J. W. and Wagner, G. (1992b). Biochemistry **31**, 8571–8586.

Perutz, M. F. (1962) *Proteins and Nucleic Acids*, Elsevier, Amsterdam.

Powers, R., Clore, G. M., Stahl, S. HJ., Wingfield, P.T. and Gronenborn, A. (1992). Biochemistry **31**, 9150–9157.

Rasmussen , B. F., Stock, A. M., Ringe, D. and Petsko, G. A. (1992). Nature **357**, 423–424.

Rosenberg, A. and Enberg, J. (1969). J. Biol. Chem. **244**, 6153–6159.

Spera, S., Ikura, M. and Bax, A. (1991). J. Biomol. NMR **1**, 155–165.

Spolar, R. S. and Record, M. T., Jr. (1994). Science **263**, 777–784.

Wittebort, R. J. and Szabo, A. (1978). J. Chem. Phys. **69**, 1723–1736.

Woessner, D. E. (1962a). J. Chem. Phys. **36**, 1–4.

Woessner, D. E. (1962b). J. Chem. Phys. **37**, 647–654.

Woodward, C. K. and Hilton, B. D. (1979). Annu. Rev. Biophys. BioEng. **8**, 99–127.

Zhang, H., Zhao, D., Revington, M., Lee, W., Jia, X., Arrowsmith, C. and Jardetzky, O. (1994). J. Mol. Biol. **238**, 592–614.

Zhao, D. and Jardetzky, O. (1994). In: *Abstracts of Papers (208th Nat'l Meeting of the Am. Chem. Soc.)*, American Chemical Society, Washington, DC, PHYS 237.

Zhao, D. and Jardetzky, O. (1995). In: *Adaptation of Simulated Annealing to Chemical Optimization Problems*, (J. H. Kalivas, ed.), Elsevier Science B.V., Amsterdam, pp. 303–328.

Zheng, Z., Czaplicki, J. and Jardetzky, O. (1995a). Biochemistry **34**, 5121–5223.

Zheng, Z., Gryk, M. R., Finucane, M. D. and Jardetzky, O. (1995b). J. Magn. Reson., Ser. B **108**, 220–234.

9

THE INVESTIGATION OF PROTEIN DYNAMICS VIA THE SPECTRAL DENSITY FUNCTION

R. Andrew Atkinson* and Jean-François Lefèvre†

UPR 9003 du CNRS
Ecole Supérieure de Biotechnologie de Strasbourg
Bld. Sébastien Brant, 67400 Illkirch, France

ABSTRACT

Spectral density mapping provides direct access to information on protein dynamics, by the solution of the equations for heteronuclear relaxation rates, with no assumptions as to the nature of the molecule or its dynamic behaviour. Reduced spectral density mapping allows the detailed characterisation of a protein's motions at a lower experimental burden than the full approach. Spectral density profiles may readily be interpreted to yield information on flexibility or slow conformational exchange and to describe the anisotropic nature of the molecule.

1. INTRODUCTION

An understanding of the dynamic behaviour of biological macromolecules is essential to complement our knowledge of structure and to provide a means to comprehend function. Much of a molecule's function may be understood on the basis of structure alone, but a static picture is not a truly satisfactory means with which to fully explain the behaviour of biological systems. Furthermore, a great deal of important processes, such as folding and many intermolecular interactions, rely on the properties of molecules or parts thereof that are difficult to characterise structurally, precisely because they lack defined structure. While certain molecular motions may be irrelevant, the utility of flexibility or conformational exchange, the changes occurring on interaction or the behaviour of folding intermediates are problems that are more difficult to address than structure *per se*. Of

---

* Present address: National Institute for Medical Research, The Ridgeway, Mill Hill, London, NW1 1AA, U.K.
† Author to whom correspondence should be addressed.

Protein Dynamics, Function, and Design, edited by Jardetzky *et al.*
Plenum Press, New York, 1998

course, a certain amount of dynamic information is reflected in the inability to define structure, be it in high crystallographic temperature factors or large r.m.s.d. values between refined NMR structures, but it is rather unsatisfactory to rely on the shortcomings of structural methods to gain an insight into dynamic behaviour.

At present, ^{15}N and ^{13}C relaxation rates are commonly analysed using the so-called 'model-free' approach, described by Lipari and Szabo (1982a,b), in which the equations for the relaxation rates (Abragam, 1961) are fitted, using three parameters, namely the order parameter, S^2, an overall correlation time, τ_c, and a correlation time for internal motion, τ_i. A fourth parameter, R_{ex}, an exchange rate term, may be added to the transverse relaxation rate, if required for a satisfactory fit, and τ_c may be left free to account for anisotropy (Phan et al., 1996). The drawbacks of this approach are becoming apparent, in particular the loss of information on nanosecond time-scale motions (Korzhnev et al., 1997) and the difficulties in interpreting data from anisotropic molecules (Luginbühl et al., 1997). Perhaps the greatest drawback is its lack of transparence and the requirement of testing a number of models to fit the data satisfactorily.

Spectral density mapping has been proposed as a more straightforward manner in which to analyse relaxation data (Peng & Wagner, 1992a; Peng & Wagner, 1992b; Farrow et al., 1995a; Ishima & Nagayama, 1995a; Peng & Wagner, 1995; Dayie et al., 1996), requiring no assumptions and allowing a simple reading of the molecule's dynamic behaviour. Indeed, on passing from relaxation rates to spectral density values, no assumptions as to the molecule's shape or motions are made, while the information we require on the dynamics of the system are readily apparent. Models are, of course, then necessary to extract quantitative information, but not to interpret the data in terms of flexibility, conformational exchange and anisotropy, or to compare the behaviour of related systems. Although simple to implement and interpret, the approach has not been widely used to date (Farrow et al., 1995b; Ishima & Nagayama, 1995b; Peng & Wagner, 1995; Lefèvre et al., 1996; Mer et al., 1996).

2. BASIC EQUATIONS DESCRIBING MOLECULAR MOTIONS IN SOLUTION

2.1. Expressions for C(T) and J(ω)

Using NMR relaxation, we may observe the motion of X-^{1}H vectors along the peptide chain of a protein in solution. X may be either the ^{15}N nucleus of the amide group or the ^{13}C nucleus of the alpha carbon. This relaxation corresponds to the evolution of the spin system (X-^{1}H) back towards its equilibrium state via transitions between the four energy levels (Fig. 1). These transitions are due to fluctuations of the local magnetic field, B_L, which arises mainly from dipolar interactions between X and ^{1}H, and the chemical shift anisotropy of X. The magnetic fields produced by these two sources vary with the orientation of the X-^{1}H vector.

The random motion of an X-^{1}H vector is composed of the overall tumbling of the molecule and local fluctuations of the structure. Each component of the motion follows a random function, $P_i(t)$, describing the variation of the orientation with time. $P_i(t)$ is a Wigner matrix with components equivalent to the second rank spherical harmonics, Y_{2q} $(\Omega_n(t))$ (Woessner, 1962a, b). It has direct physical effects that may be read via its two fundamental functions: the correlation function, $C_i(t)$, and the spectral density function, $J_i(\omega)$. $C_i(t)$ describes the mean time evolution:

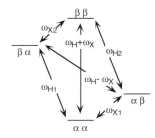

Figure 1. Four energy level diagram for a two spin system H - X. ω_{H1}, ω_{H2} and ω_{X1}, ω_{X2} correspond to simple quantum transitions at lower (1) and upper (2) frequency components of the doublets. $(\omega_H - \omega_X)$ and $(\omega_H + \omega_X)$ are the zero and double quantum transitions respectively. In the case of ^{15}N, the lower energy level being β, the energy levels should be permuted : $\alpha\alpha$ with $\alpha\beta$ and $\beta\alpha$ with $\beta\beta$.

$$C_i(\tau) = <P_i(t) . P_i(t + \tau)> \tag{1}$$

$J_i(\omega)$ describes this motion in the frequency domain, the inverse of the time domain. The two functions are related by a Fourier Transform:

$$J_i(\omega) = \int_{-\infty}^{+\infty} C_i(\tau) . \exp(i.\omega.\tau) . d\tau \tag{2}$$

NMR relaxation samples the motion through $J_i(\omega)$ which gives the amplitudes of the motions at a given set of frequencies: 0, ω_X, $\omega_H - \omega_X$, ω_H and $\omega_H + \omega_X$.

Most often, the various motions act independently. However, if two motions are not independent, they are each described by their auto-correlation function (Eq. 1) and their cross-correlation functions:

$$C_{ij}(\tau) = <P_i(t) . P_j(t + \tau)> \tag{3}$$

These cross-correlated motions act at very similar frequencies and cannot be separated within the observations. In the following, we will only consider non-correlated motions.

The global correlation function is given by the product of the contributions of the individual motions:

$$C(t) = \prod_i C_i(t) \tag{4}$$

The overall motion depends on the shape of the molecule. If the molecule, hydrated, is totally spherical, the X-^{1}H vectors experience its isotropic motion and the time dependence of $C_0(t)$ is a simple exponential while its spectral density function is a Lorentzian (see 2.2 below for the case an anisotropic molecule):

$$C_0(t) = \frac{1}{5} . \exp(- \frac{t}{\tau_c})$$

$$J_0(\omega) = \frac{2}{5} . \frac{\tau_c}{1 + \omega^2.\tau_c^2} \tag{5}$$

with a correlation time τ_c:

$$\tau_c = \frac{1}{6.R} \tag{6}$$

R being the constant of rotational diffusion:

$$R = \frac{k.T}{8.\pi.\eta.r^3} \tag{7}$$

where η is the viscosity of the medium and r is the radius of the hydrated sphere. Because the molecule may rotate freely about any axis, the asymptotic value of $C_o(t)$ decreases to zero as t goes to infinity. In other words, there is no correlation between the orientation of the X-^{1}H vectors in the molecule at the starting point and after a very long delay. This is not the case for the internal motions. The space domain explored by the X-^{1}H vector moving inside the structure, is limited by energy barriers. The correlation function therefore contains a non-zero asymptotic value:

$$C_i(t) = (1 - S_i^2) \cdot \exp(-\frac{t}{\tau_i}) + S_i^2 \tag{8}$$

S_i^2 is called the order parameter attached to the motion i and corresponds to the asymptotic value of $C_i(t)$. The global correlation function becomes a sum:

$$C(t) = \frac{1}{5} \cdot \left\{ S^2 \cdot \exp(-\frac{t}{\tau_c}) + \sum_j S_j^2 \cdot \exp(-\frac{t}{\tau_j}) \right\} \tag{9}$$

where S^2 and S_j^2 are defined by:

$$S^2 = \prod_j S_j^2 \quad \text{and} \quad S_j^2 = \prod_{k,l} (1 - S_k^2) \cdot S_l^2 \tag{10}$$

where k and l vary over all values of i, and k is not equal to l. The correlation times are given by the various sums:

$$\frac{1}{\tau_j} = \frac{1}{\tau_c} + \sum_k \frac{1}{\tau_k} \tag{11}$$

The spectral density function is also obtained as a sum of various components:

$$J(\omega) = \frac{2}{5} \cdot \left\{ S^2 \cdot \frac{\tau_c}{1 + \omega^2.\tau_c} + \sum_j S_j^2 \cdot \frac{\tau_j}{1 + \omega^2.\tau_j} \right\} \tag{12}$$

After attempts to use various models of motions (for review, see Dayie et al. 1996), it was proposed to extend the sum in the equations of C(t) or J(ω) progressively, to fit the NMR relaxation data (King & Jardetzky, 1978). It was then suggested that the contribution of the internal motions might be simplified with a unique $C_i(t)$ (Lipari & Szabo, 1982a,b). This leads to the now familiar expressions for $C(t)$ and $J(\omega)$:

$$C(t) = S^2 . \exp(-\frac{t}{\tau_c}) + (1 - S^2) . \exp(-\frac{t}{\tau_e})$$

$$\text{with} \quad \frac{1}{\tau_e} = \frac{1}{\tau_c} + \frac{1}{\tau_i}$$

$$J(\omega) = \frac{2}{5} . \left\{ S^2 . \frac{\tau_c}{1 + \omega^2 . \tau_c} + (1 - S^2) . \frac{\tau_e}{1 + \omega^2 . \tau_e} \right\} \tag{13}$$

More complicated models have been elaborated containing two types of internal motion, one 'slow' (ps-ns) and the other 'fast' (< ps) (Clore et al., 1990). A slow exchange contribution (in the µs-ms range) is also often added to the value of $J(0)$. All these parameters are adjusted to fit three experimental NMR observations: the longitudinal ($R_X(X_Z)$), the transversal ($R_X(X_X)$) and the cross ($R_X(H_Z \rightarrow X_Z)$) relaxation rates.

2.2. The Case of Molecular Anisotropy

In an ellipsoidal molecule, rotational motions about its three axes occur at different rates. Proteins are often found to approximate to axially symmetric ellipsoids. Rotation about the principal axis is faster than that about the axes orthogonal to it, such that the orientation of X-^{1}H vectors with respect to the axes affects their motions and thus determines the manner in which the X nucleus is relaxed. For example, the X nucleus in a vector aligned with the long axis is not relaxed by the fast rotation about that axis, but only by the slower rotation about the other axes; conversely, an X nucleus in a vector perpendicular to the long axis is relaxed principally by rapid re-orientation about that axis. The effect is that the correlation function and the spectral density values along the sequence of an anisotropic molecule depend on the orientation of the X-^{1}H vector with respect to the axes.

The fluctuating rotations around each axis follow correlation functions ($C_{oi}(t)$) which evolve to non-zero values, in an analogous manner to $C_i(t)$, expressed in Eq. 8. This arises from the fact that each of these rotations maintains the vector on a cone around an axis. The various components, $C_{oi}(t)$, must be treated to account for the rotation of each axis in the laboratory co-ordinate frame. The product of correlation functions, which leads to the overall correlation function corresponding to molecular tumbling, may be expressed as a sum, as may $J(\omega)$ (Woessner, 1962a,b). In the simpler case of an axially symmetric ellipsoid, these sums have three components:

$$C_o(t) = \frac{1}{5} . \sum_{i=1->3} A_i . \exp(-\frac{t}{\tau_{ci}})$$

$$J_o(\omega) = \frac{2}{5} . \sum_{i=1->3} A_i . \frac{\tau_i}{(1 + \omega^2 . \tau_{ci}^2)} \tag{14}$$

where the coefficients A_i depend on the angle, θ, between the X-^{1}H vector and the principal axis:

$$A_1 = \tfrac{1}{4} . (3.\cos^2\theta - 1)^2 \quad , \quad A_2 = 3.\sin^2\theta.\cos^2\theta \quad , \quad A_3 = \tfrac{3}{4}.\sin^4\theta \tag{15}$$

The three correlation times describing the motion of the ellipsoid may be determined using:

$$\frac{1}{\tau_{c1}} = 6.R_b \quad , \quad \frac{1}{\tau_{c2}} = R_a + 5.R_b \quad , \quad \frac{1}{\tau_{c3}} = 4.R_a + 2.R_b \tag{16}$$

Where R_x are the rotational diffusion constants. Given the half-axes a, b and c of an axially symmetric prolate ellipsoid (where b=c and a>b):

$$R_a = \frac{3.k.T.\left(2.a - b^2.S\right)}{32.\pi.\eta.b^2.\left(a^2 - b^2\right)} \quad , \quad R_b = R_c = \frac{3.k.T.\left[\left(2a^2 - b^2\right).S - 2.a\right]}{32.\pi.\eta.\left(a^4 - b^4\right)} \tag{17}$$

where:

$$S = 2.\left(a^2 - b^2\right)^{-\frac{1}{2}}.\ln\left[\left(a + \left(a^2 - b^2\right)^{\frac{1}{2}}\right)/b\right]$$

The simple equation given for spherical molecules (Eq. 5) can be confirmed. The fact that a = b gives : $R_a = R_b = R_c = R$ (Eq. 7), and $\tau_{c1} = \tau_{c2} = \tau_{c3} = \tau_c$ (Eq. 6). The sum of the coefficients, A_i, is always equal to one. This leads from Eq. 14 to Eq. 5.

For molecules that are not axially symmetric, the motions involve five correlation times (τ_{ci}) and require the solution of a set of elliptic integrals, but the spectral density may also be calculated, if rather laboriously.

3. CALCULATION OF THE SPECTRAL DENSITY, J(Ω)

3.1. Full Spectral Density Mapping

Spectral density mapping simply involves solving Abragam's equations for heteronuclear relaxation rates (Abragam, 1961), expressed in terms of the spectral density, J(ω). In the full approach (Peng & Wagner, 1992a; Peng & Wagner, 1992b, Peng & Wagner, 1995), a set of six independent relaxation rate measurements ($R_X(X_Z)$, $R_X(X_X)$, $R_X(H_Z \rightarrow X_Z)$, $R_X(2X_ZH_Z)$, $R_X(2X_XH_Z)$, $R_H(H_Z)$) is required to determine the spectral density, J(ω) at the five frequencies appearing in the equations (0, ω_X, $\omega_H - \omega_X$, ω_H, $\omega_H + \omega_X$), and the dipolar term for the relaxation of the amide proton by other protons, $\rho_{HH'}$. The equations may be expressed most simply in matrix form:

$$\begin{bmatrix} R_X(X_Z) \\ R_X(X_X) \\ R_X(H_Z \rightarrow X_Z) \\ R_X(2X_ZH_Z) \\ R_X(2X_XH_Z) \\ R_H(H_Z) \end{bmatrix} = \begin{bmatrix} 0 & E & A & 0 & 6A & 0 \\ 2E/3 & E/2 & A/2 & 3A & 3A & 0 \\ 0 & 0 & -A & 0 & 6A & 0 \\ 0 & E & 0 & 3A & 0 & 1 \\ 2E/3 & E/2 & A/2 & 0 & 3A & 1 \\ 0 & 0 & A & 3A & 6A & 1 \end{bmatrix} \times \begin{bmatrix} J(0) \\ J(\omega_X) \\ J(\omega_H - \omega_X) \\ J(\omega_H) \\ J(\omega_H + \omega_X) \\ \rho_{HH'} \end{bmatrix} \tag{18}$$

or R = C × J, where:

$$A = \left(\frac{\mu_0}{4.\pi} \right)^2 . \frac{\gamma_H^2.\gamma_X^2.h^2}{4.r_{XH}^6} \quad , \quad B = \frac{\Delta_X^2.\omega_X^2}{3} \quad , \quad E = 3.A + B \tag{19}$$

where μ_0 is the permeability of the vacuum, γ_H and γ_X are the gyromagnetic ratios of the ^{1}H and X nuclei respectively, r_{XH} is the internuclear X-^{1}H bond distance and Δ_X is the chemical shift anisotropy.

The values of the spectral density, J(ω), may thus be obtained by solving the equations, or by inverting the matrix **C** and applying:

$$J = C^{-1} \times R \tag{20}$$

3.2. Reduced Spectral Density Mapping

A more practical approach in experimental terms, referred to as 'reduced' spectral density mapping (Farrow et al., 1995a; Ishima & Nagayama, 1995a; Peng & Wagner, 1995; Lefèvre et al., 1996), requires only three relaxation rates, $R_X(X_Z)$, $R_X(X_X)$, and $R_X(H_Z \rightarrow X_Z)$, corresponding to the measurement of T_1, T_2 and the heteronuclear NOE, respectively. It is assumed that the spectral density function is flat around ω_H (Ishima & Nagayama, 1995a; Lefèvre et al., 1996), allowing the equations for the relaxation rates to be simplified by setting $J(\omega_H-\omega_X) = J(\omega_H) = J(\omega_H+\omega_X)$. The equations expressed in matrix form then simplify to:

$$\begin{bmatrix} R_X(X_Z) \\ R_X(X_X) \\ R_X(H_Z \rightarrow X_Z) \end{bmatrix} = \begin{bmatrix} 0 & E & 7A \\ 2E/3 & E/2 & 13A/2 \\ 0 & 0 & 5A \end{bmatrix} \times \begin{bmatrix} J(0) \\ J(\omega_X) \\ \langle J(\omega_H) \rangle \end{bmatrix} \tag{21}$$

Farrow et al. (1995a) have proposed that $J(0.87.\omega_H)$ may be determined from $R_X(X_Z)$ and $R_X(H_Z \rightarrow X_Z)$ alone, from which $J(0)$ and $J(\omega_N)$ may be calculated. Ishima and Nagayama (1995a), on the other hand, show that $<J(\omega_H)>$ in Eq.(3) may be more correctly referred to as $J(\omega_H+\omega_X)$.

3.3. Applicability to ^{13}C Relaxation

We have shown through model calculations (Atkinson & Lefèvre, 1998), that, for ^{15}N relaxation data, the simplification used in reduced spectral density mapping is a good approximation to the true spectral density. For ^{13}C relaxation data, however, the approximation that $J(\omega_H)$ is flat at ω_H is less valid, due to the higher frequency of the ^{13}C nucleus compared to ^{15}N. A second approach to the calculation of the spectral density from the relaxation rates involves the fitting of the spectral density function around ω_H by a single Lorentzian, characterised by a correlation time, τ_H. The results of the two approaches serve to delimit the range of values of J(ω) within which the true value lies.

4. SIMPLE GUIDE TO THE INTERPRETATION OF SPECTRAL DENSITY

The greatest advantage of the use of spectral density mapping for the study of protein dynamics is that the relaxation rates are interpreted making no assumptions as to the nature of the molecule or its motions. The spectral density profiles at frequencies 0, ω_X and ω_H are only interpreted subsequently with criteria appropriate to the system under study. Here, we shall describe the straightforward analysis of spectral densities in six points. Below, we will expand upon the effects of differing motions on the spectral density, the nature of exchange contributions and the impact of anisotropy on the profiles. This will be illustrated with examples taken from the literature and we will refer back to these six points in the text below.

§1. The baseline of the spectral density profiles represents the global behaviour of the molecule.

§2. Decreases in $J(0)$, accompanied by rises in $J(\omega_H)$ and fluctuation of $J(\omega_X)$, indicate regions of greater flexibility (but beware the effects of anisotropy, below).

§3. For spherical molecules, the correlation times for overall and internal motions are given by the points at which a linear fit of $J(\omega_X)$ *vs.* $J(0)$, for the ensemble of X-^{1}H vectors, intercepts the theoretical curve for isotropic motion (see effects of anisotropy, below).

§4. Sudden changes in $J(0)$, accompanied by fluctuations of $J(\omega_X)$ and $J(\omega_H)$ may result from the anisotropic nature of the molecule. (In such a case, the baseline represents the orientation of X-^{1}H vectors with respect to the principal axis most represented in the molecule, generally the elements of secondary structure).

§5. For an anisotropic molecule, the linear fit of $J(\omega_X)$ *vs.* $J(0)$ for the ensemble of X-^{1}H vectors will be weighted toward the orientation most represented in the molecule. The correlation times obtained may not readily be related to the physical dimensions of the molecule. More sophisticated fitting procedures may be used to obtain the lengths of the axes and the orientations of the X-^{1}H vectors.

§6. Increases in $J(0)$, with no corresponding change in $J(\omega_X)$ and $J(\omega_H)$ indicate slow conformational exchange.

5. ANALYSIS OF THE SPECTRAL DENSITY PROFILES OF SPHERICAL PROTEINS

5.1. Rigid Sphere

For an X-^{1}H bond vector in a rigid sphere, its relaxation may be described by a single correlation time, τ_c, for the overall tumbling of the sphere in solution. This correlation time is given by Eq. 6, and the spectral density, $J(\omega)$, then has the form of a simple Lorentzian (Eq. 5). For an improbable protein that behaves as a rigid sphere with a given radius, the relaxation of its X nuclei will be governed by the spectral density values, $J(\omega)$, at the set of relevant frequencies (0, ω_X, $\omega_H-\omega_X$, ω_H, $\omega_H+\omega_X$), appearing in the equations of Abragam (1961). These, of course, will give uniformly flat profiles when plotted against sequence (§1).

Figure 2. Plot showing the theoretical spectral density at frequencies 0 and ω_X. The curve represents the values for a rigid sphere as τ_c is increased. The straight line indicates variation of the relative weightings of the overall and internal motions. Circles denote the spectral density for X-^{1}H vectors in an anisotropic molecule as the angle, θ, to the principal axis is varied. Unit values are not given on axes as they depend on the used resonance frequency.

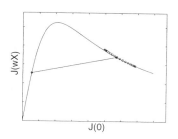

A useful tool for interpreting data, mentioned above, is the plot of $J(\omega_X)$ *vs.* $J(0)$ which, for a rigid sphere undergoing isotropic tumbling, gives a point on the theoretical curve corresponding to τ_c (Fig. 2). Its position, under given experimental conditions, depends only on the radius of the hydrated protein (§3, Eq. 6, 7). This curve of $J(\omega_X)$ *vs.* $J(0)$, as shown in Fig. 2, goes through a maximum occurring for the correlation time value (τ_m) which depends on the type of observed nucleus and the amplitude of the magnetic field. Expressed as a function of proton resonance frequency (ν_H), $\tau_m = 1.57/\nu_H$ for ^{15}N and $0.63/\nu H$ for 13C (3.14ns and 1.27ns respectively if $\nu_H = 500$MHz). For proteins, the overall correlation time is larger than the given values of τ_m.

5.2. Internal Motion Faster than Overall Correlation Time τ_c

An uncorrelated and faster motion of a X-^{1}H vector, within the tumbling sphere, characterised by a correlation time τ_i, will affect the spectral density according to Eq. 12 (13 if it is a unique internal motion). The effect on the spectral density of this additional motion is to reduce $J(0)$ from its rigid sphere value and raise $J(\omega_H+\omega_X)$, with the effect on $J(\omega_X)$ depending on ω_X, the correlation times of the motions, τ_c and τ_i, and its weighting, $(1 - S_i^2)$ (Eq. 8 and see above, 5.1).

For a spherical protein in which all X-^{1}H vectors are relaxed by similar motions, the spectral density profiles along the sequence will be altered but remain flat (§1). In the plot of $J(\omega_X)$ *vs.* $J(0)$, the spectral densities will fall at a single point. Now, if the internal motions have maximal amplitudes $(S_i^2 = 0)$, this point will again lie on the theoretical curve $J(\omega_X)$ *vs.* $J(0)$, corresponding to a mixture between the various correlation times (Eq. 11, 13). Physically, $S_i^2 = 0$ means that the positions of the vectors within the sphere have no restrictive constraints. If $S_i^2 \neq 0$ the point falls between the values on the theoretical curve corresponding to τ_c and τ_j (Eq. 12 and 13). If these two correlation times are known, the order parameter which modulates the overall tumbling motion, S^2 (Eq.13), can be deduced.

5.3. Example: Calcium-Loaded Calbindin D9k

Variability in the contributions of internal motions to the spectral density will lead to fluctuations of $J(\omega)$ from the baseline behaviour (§2). A good case of a spherical protein is given by the calcium-loaded form of calbindin D9k (Kördel et al., 1992). The molecule is close to spherical, with half-axis lengths of 15.5 Å, 14.4 Å and 12.9 Å. The spectral density profiles reveal flat plots at all three frequencies, all along the sequence (§1), except at the N- and C-termini, and a central loop connecting the two EF-hands (Fig. 3). These portions of the molecule show greater flexibility than the core of the protein (§2). This led to an inability to define well the structure for these residues.

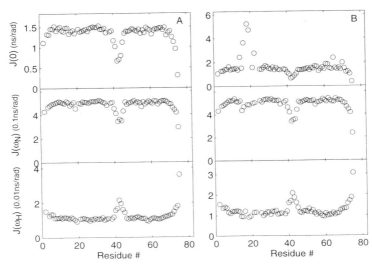

Figure 3. Plot of J(ω) *vs.* sequence for the calcium-binding protein, calbindin D$_{9k}$, with both sites occupied by Ca^{2+} (A) and with one site occupied by Cd^{2+} (B). The ^{15}N relaxation data used to calculate J(ω) are from Kördel et al. (1992) and Akke et al. (1993).

In the plot of J(ω$_X$) *vs.* J(0), differences in the time-scales of the internal motions for each residue would lead to points lying on lines with different slopes. In fact, the plot of the calbindin data (Fig. 4), shows that the points lie neatly along a straight line. A simple fit shows that this line intercepts the theoretical curve for a rigid spherical molecule at two points (§3). The spectral density values at these points correspond to correlation times of 4.58 ns and 0.87 ns. The longer correlation time is clearly τ$_c$ while the shorter is τ$_e$ (§3). Assuming a spherical model, the overall correlation time of 4.58 ns corresponds to a molecular radius of 16.2 Å, slightly larger than the longest half-axis of the molecule. This is often observed and arises from the hydration of the protein.

The points at which the linear fit to the ensemble of data points intercepts the theoretical curve may be determined also using a simple mathematical method based on a third order equation (Dayie et al., 1995, Lefèvre et al., 1996, Dayie et al., 1996). This yields three correlation times, one of which has no physical meaning. It is often negative, though for large molecules it may be positive but correspond to an unrealistically short overall correlation time. The remaining two points yield the correlation times τ$_c$ and τ$_e$, and all experimental points should lie between these two points.

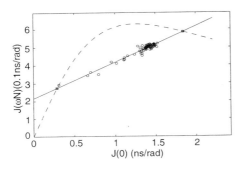

Figure 4. Plot of J(ω$_N$) *vs.* J(0) for calcium-loaded calbindin D$_{9k}$. (—) linear fit of the experimental data (o). (--) theoretical values of J(ω$_N$) *vs.* J(0) for varying correlation times. (⊗) points of intersection between the experimental fit and the theoretical curve giving : τ$_c$=4.58ns and τ$_c$=0.87ns.

The behaviour of $J(\omega_H)$ along the sequence is always inverted with respect of that of $J(0)$. $J(\omega_X)$ behaves according to the slope of its linear fit observed with $J(0)$ (Fig. 4). If this slope is positive or negative, $J(\omega_X)$ moves according to or against the fluctuations of $J(0)$. Of course, $J(\omega_X)$ may also be not moving at all (see Fig. 10 for example).

It is very interesting to note that all X-^{1}H vectors appear to experience internal motions with an identical fluctuating correlation time, characterised by τ_e. This simply indicates that each residue is sensitive to the motions affecting its neighbours' positions. Consequently, the X-^{1}H vector of each residue experiences the same set of motions, with an amplitude modulated by its environment. The peptide chain is agitated by motions transmitted from one end to the other.

The behaviour of $J(\omega)$ along the sequence appears as a continuous function. At each frequency, $J(\omega)$ progressively changes as the sequence evolves from a rigid to mobile region or the inverse. This reflects increases and decreases in the amplitude of the motion on passing through a loop or hinge region.

The correlation times read from the reduced spectral density function behaviour may be compared to those obtained (Kördel et al., 1992) using the so called 'model-free' formalism (Lipari and Szabo 1982a,b): the overall correlation time was found to be 4.25 ns and the correlation times for internal motions lie between 0.0 ps and 61.1 ps (*much* shorter than the value of 0.87 ns given above), except for residues in the central loop (Lys41-Ser44) and the C-terminus (Ser74-Gln75). For these six residues, the 'extended model-free' formalism (Clore et al., 1990) was applied and gives correlation times for internal motions greater than 1 ns. The main difference between the two methods (Fig. 5) arises from the fact that, in the reading of the spectral density function, the relationship between the $J(\omega)$ of all residues is used. Once the overall and internal correlation times are known, the calculation proposed in the 'model-free' approach and given in Eq. 13 can be applied to get the values of S^2.

6. ANALYSIS OF THE SPECTRAL DENSITY PROFILES OF ANISOTROPIC PROTEINS

6.1. Molecular Anisotropy

We may first consider the effects of molecular anisotropy, in an axially symmetric ellipsoid, in the absence of internal motions. As the angle to the principal axis, θ, is varied, the contributions of movements about the axes to the spectral density are altered. On moving from $\theta=0°$ to $\theta=90°$, faster motions about the principal axis are introduced into the set

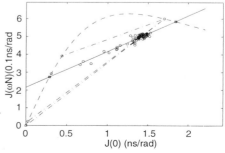

Figure 5. Plot of $J(\omega_N)$ *vs.* $J(0)$ for calcium-loaded calbindin D_{9k}. Comparison between the (—) linear fit of the experimental data (o) (as in Fig. 5) and results of the 'model-free' approach (-··-).

of motions re-orienting the X-^{1}H vectors, leading to a decrease in J(0) and corresponding changes in $J(\omega_X)$ and $J(\omega_H)$ (Eq. 14). In the absence of internal motion, the plots of $J(\omega_X)$ vs. J(0) depart from the theoretical curve for isotropic tumbling to an extent that depends on the degree of anisotropy (Fig. 2). The slope of $J(\omega_X)$ vs. J(0) depends on the domain of correlation times with respect to τ_m for which occurs the maximum value of the theoretical curve shown in Fig. 2. For proteins, this slope will always be negative. For vectors strongly moved by the overall tumbling, $J(\omega_X)$ will decrease when J(0) increase and inversely. This aspect will be useful in reading the behaviour of $J(\omega)$ along the sequence.

Internal motions may be introduced, as for molecules undergoing isotropic tumbling (Zheng et al., 1995), but this assumes that the anisotropy has no effect on the contribution of the internal motion, and results in similar spectral density values when the overall motions contribute little, whatever the orientation of the X-^{1}H vector with respect to the principal axis. This is not confirmed by reading of experimental spectral density data.

6.2. Example: eglin c

A detailed analysis of the behaviour of J(0) along the sequence of eglin c (Fig. 6) already suggests that this protein is ellipsoidal. In the sequence of the rigid part of the protein, J(0) appears to fluctuate abruptly from one residue to the other, with a larger amplitude than that due to experimental error. The plot of $J(\omega_N)$ vs. J(0) shows also that the experimental points often lie far from the linear fit. This is clearly due to the anisotropy of the protein. The X-^{1}H vectors experience different overall correlation times, depending on their orientation. Their internal motions characterised by τ_e also depend on this orientation.

As shown in Fig. 7, upper and lower limits for the data can be found easily, giving the overall and internal correlation times for the principal and perpendicular axes. The evolution of these correlation times with the angle to the principal axis, θ, are given in Eq. 16. A simple grid search allows θ to be determined for each ^{15}N-^{1}H vector. These orientation parameters are compared in Fig. 8 with those obtained directly from the three-dimensional structures solved by crystallography (Hipler et al., 1992 and NMR (Hyberts et al., 1992). As can be seen, the orientations of the ^{15}N-^{1}H vectors sometime differ between the

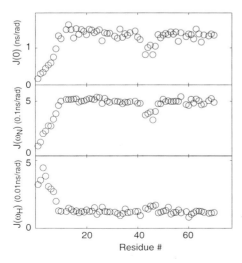

Figure 6. Plot of the spectral density function values at 0, ω_N and ω_H angular frequencies, vs. the sequence of the serine protease inhibitor eglin c. The N-terminal sequence appears unstructured by both NMR and X-ray. The loop (41–46) has a well-defined structure but moves with respect to the core of the protein (Hyberts et al., 1992; Peng & Wagner, 1995).

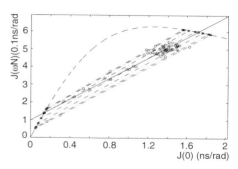

Figure 7. Plot of $J(\omega_N)$ *vs.* $J(0)$ for eglin c.(—) Line fitting the data as in a spherical case. (-··-) A fit with the two borders of the data points have defined the correlation attached to the larger axis (down border, $\tau_c = 4.54$ns and $\tau_e = 0.13$ns) and the smaller one (up border, $\tau_c = 3.9$ns and $\tau_e = 0.46$ns). The lines plotted within the data, between the two borders, correspond to increasing values of Θ. The increment $\Delta\Theta$ is equal to 15 deg. in order to get a readable figure.

X-ray and NMR structures, while those extracted from the relaxation data fit with one or the other set of structural results.

7. SLOW CONFORMATIONAL EXCHANGE

7.1. Effects on the Spectral Density Function

X-^{1}H bond vectors in peptide sequences undergoing slow (μs-ms) conformational exchange will be relaxed by an additional mechanism, apparent in $R_X(X_X)$, or T_2, measurements. The magnitude of this exchange term is given by:

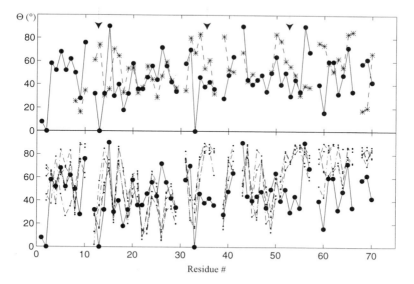

Figure 8. Plot of Θ *vs.* sequence for eglin c. Θ is the angle of orientation of ^{15}N-H vectors relative to the major axis of the ellipsoid shaping the protein. The black points in both figures represent the values of Θ obtained from $J(\omega)$ analysis (see Fig. 7, done with an increment $\Theta = 2$ deg.). (A) Comparison with the values obtained from the X-ray structure (Hipler et al., 1992). (B) Comparison with the values deduced from five three dimensional structures given by NMR (Hyberts et al., 1992). The arrows on the top show three regions where values of X-ray and NMR Θ do not fit, while the values obtained from $J(\omega)$ fit better with one or the other.

$$R_{ex} = \frac{4.\pi^2.\Delta\upsilon^2.p_A.p_B.\tau_{ex}}{\left(1+\omega_1^2.\tau_{ex}^2\right)} \tag{22}$$

where $\Delta\nu$ is the chemical shift difference between states A and B, p_A and p_B are their populations, ω_1 is the frequency of the spin-lock used and τ_{ex} is the exchange time. Because this type of motion occurs at low frequency, it results *solely* in an increase in J(0) for the X-^{1}H bond vectors involved:

$$J_{obs}(0) = J(0) + \frac{3}{2} \cdot \frac{R_{ex}}{E} \tag{23}$$

where $J_{obs}(0)$ corresponds to the experimental value of the spectral density function, and J(0) includes all motions except the slow exchange (Eq. 12, 13) and E contains the dipolar and CSA constants which are used in the calculation of $J(\omega)$ (Eq. 19).

7.2. Example: Calbindin D9k-Cd^{2+}

As shown in Fig. 3, conformational exchange is clearly in evidence for calbindin in which one of the two calcium-binding sites is occupied by a Cd^{2+} ion, while the other is vacant (Akke et al., 1993). The backbone of the occupied site is rigidified compared to the apo form of the protein (§2), while the unoccupied site clearly exhibits conformational exchange (§6). The values of J(0) arising from overall and faster internal motion and ignoring the slow exchange contribution can be deduced from the values of $J(\omega_N)$ and the plot which gives τ_c and τ_e (as in Fig. 2 and 4). The R_{ex} values are then given by to the difference between the experimental and the calculated values of J(0), using Eq. 23. The data points of residues affected by conformational exchange are shifted to higher values of J(0) in the plot of $J(\omega_N)$ *vs.* J(0) and may lie beyond the point on the theoretical curve corresponding to the overall correlation time. Once identified, these points should be excluded from the linear fit to the ensemble of data points.

8. ANISOTROPY AND SLOW CONFORMATIONAL EXCHANGE: AN IMMUNOGLOBULIN DOMAIN FROM TITIN

The effects of the anisotropic nature of a molecule on its relaxation are clearly illustrated in the case of the I28 domain of the muscle protein, titin (Improta et al., 1998). This Ig domain is predominantly β-sheet and approximates to an axially symmetric cylinder, with the principal axis twice as long as the other two axes (lengths of half-axes: 24.1 Å, 12.4 Å, 11.9 Å). The plots of spectral density *vs.* sequence (Fig. 9) show a baseline from which certain sets of residues depart (§1). Analysis of the secondary structure reveals that the baseline corresponds to ^{15}N-^{1}H vectors of residues forming the anti-parallel β-sheet structure of the domain (§4). All these vectors are close to being parallel with respect to each other, and therefore lie at approximately the same angle, almost perpendicular, to the principal axis of the domain. The ^{15}N nuclei are therefore relaxed in a similar manner, yielding the relatively flat baseline observed. Since the vectors are close to perpendicular to the principal axis, the ^{15}N nuclei of these residues are relaxed to a large extent by the fastest overall rotation about this axis, such that the J(0) values for these residues are expected to be the lowest. Fluctuations from the baseline fall into three categories:

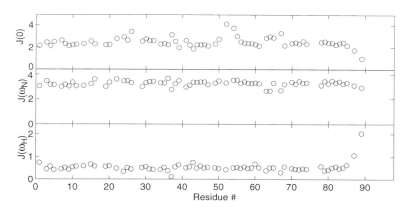

Figure 9. J(ω) extracted from the ^{15}N relaxation data *vs.* sequence of the titin I28 domain (J(0) in ns/rad, J(ω_N) in 0.1ns/rad and J(ω_H) in 0.01ns/rad). J(0) shows a low base line characterising that the majority of the ^{15}N-^{1}H vectors are in orientations perpendicular to the principal axis. The vectors for which J(ω_N) decreases while J(0) increases evoluate towards the principal axis orientation. Slow conformational exchange is occurring when J(ω_N) stays constant and does not follow the increasing fluctuation of J(0) (residues 22–26 and 50–55).

i. faster motions are observed for the C-terminal residues (§2), leading to decreased J(0) and J(ω_N) and increased J(ω_H) values, as for calcium-loaded calbindin D$_{9k}$, above;

ii. slow conformational exchange is evident for residues 22–26 and for residues 50–55 : increases in J(0) are not accompanied by fluctuations in the spectral density at higher frequencies (§6);

iii. residues for which J(0) increases while J(ω_N) drops indicate that the ^{15}N-^{1}H vector is structured but oriented such that it makes a smaller angle to the principal axis (§4). This behaviour is evident for residues 37–41 and 63–67, and is to be expected in all anisotropic molecules.

It should be noted that, if the ^{15}N-^{1}H vectors of the secondary structure of an anisotropic molecule are aligned with the principal axis, the fluctuations will be inverted: these may still be distinguished from faster motions (as in (i), above) under conditions where J(ω_N) increases,for faster motions. Furthermore, if the approximate dimensions of the hydrated molecule are known, the expected ranges of spectral density values for ^{15}N-^{1}H vectors at angles between 0° and 90° to the principal axis may be calculated: fluctuations exceeding these ranges cannot arise from anisotropy alone.

Figure 10. J(ω_N) vs. J(0) for the I28 domain of titin. The points very close to the theoretical curve (--) and on its right correspond to residues affected by slow exchange motions (see Fig.9). (—) This straight line correspond to the linear fit on the ensemble of data. The dispersion of the data around this fit reflects the anisotropy of this protein. The down and up fits (-··-) gives the correlation times attached respectively to the principal and its perpendicular axes. The line (-··-) in the middle, very similar to the global fit (—) corresponds to 45 deg angle with respect to the principal axis.

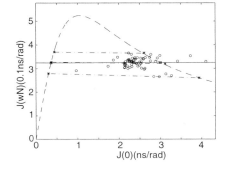

For anisotropic molecules, a fit to the ensemble of data points in the plot of $J(\omega_N)$ vs. $J(0)$ can still be used (as shown for eglin c, 6.2) to yield correlation times for the motions of the molecule. This plot for titin I28 domain also help the reading of $J(\omega)$ along the sequence. It can be easily observed in Fig. 10 that some ^{15}N-^1H exhibit too large $J(0)$. The corresponding points, $J(\omega_N)$ vs. $J(0)$, fall outside the permitted domain which is limited by the theoretical curve. It clearly indicates that slow exchange affects the behaviour of these vectors.

Of course the assumption that the molecule is spherical is not completely valid (§5). For the I28 domain of titin, such an analysis gives an overall correlation time of 8.06 ns and a correlation time for internal motions of 1.03 ns. If the molecule were spherical, the overall correlation time would correspond to a molecule of radius 19.8 Å. However, an analysis attached to the anisotropy can be done as described for eglin c above (6.2). As seen in Fig. 10, a large majority of the ^{15}N-^1H vectors appears perpendicular to the principal axis, corresponding to the observation of the three dimensional structure. The fit on both sides of the ensemble of $J(\omega_N)$ vs. $J(0)$ data gives two almost parallel lines. More experimental points are close the upper fit line than to the lower one. Fig. 10 also shows why $J(\omega_N)$ and $J(0)$ have opposite fluctuations along the sequence.

CONCLUSION

The reading of the spectral density functions is in fact quite simple. It directly gives interesting information on the shape of the protein, on the orientation of the observed vectors and their dynamic behaviour characterised by fast internal motions and sometimes slow exchange process. One of the main conclusions which appears by this reading of $J(\omega)$ is that all vectors, situated along the peptide chain, feel internal motions characterised by the same correlation time. This is certainly due to the physical contact from one residue to the other along this chain. The already simple treatment of the relaxation data called 'model-free' (Lipari and Szabo, 1982a,b) is in fact still more simplified by this method. The anisotropy of most of the protein (the sphere is almost only a theoretical model) can be carefully read through this 'J(ω) reading'.

REFERENCES

Abragam, A. (1961) "Principles of Nuclear Magnetism," Oxford University Press, Oxford.
Akke, M., Skelton, N.J., Kördel, J., Palmer III, A.G. & Chazin, W.J. (1993) *Biochemistry 32*, 9832–9844.
Atkinson, R.A. & Lefèvre, J.-F. (1998) submitted for publication.
Clore, G.M., Szabo, A., Bax, A., Kay, L.E., Driscoll, P.C., and Gronenborn, A.M. (1990) *J. Am. Chem. Soc. 112*, 4989–4991.
Dayie, K.T., Wagner, G. & Lefèvre, J.-F. (1996) *in* "Dynamics and the Problem of Recognition in Biological Macromolecules" (Jardetzky, O. & Lefèvre, J.-F., Eds.), Plenum Press, New York.Series A : Life Sciences. Vol 288, pp.139–162.
Dayie, K. T., Lefèvre, J.F. & Wagner, G. (1996) *Ann. Rev. Phys. Chem. 47*, 243–282.
Farrow, N.A., Zhang, O., Szabo, A., Torchia, D.A. & Kay, L.E. (1995a) *J. Biomol. NMR 6*, 163–162.
Farrow, N.A., Zhang, O., Forman-Kay, J.D. & Kay, L.E. (1995b) *Biochemistry 34*, 868–878.
Hipler, K. Priestle, J.P., Rahuel, J. & Grütter, M.G. (1992) *FEBS Lett. 309*, 139–145.
Hyberts, S.G., Goldberg, M.S., Havel, T.F. & Wagner, G. (1992) *Protein Sci. 1*, 736–751.
Improta, S., Krueger, J.K., Gautel, M., Atkinson, R.A., Lefèvre, J.-F., Moulton, S., Trewhella, J. & Pastore, A. (1998) submitted for publication.
Ishima, R. & Nagayama, K. (1995a) *J. Magn. Reson. Series B 108*, 73–76.
Ishima, R. & Nagayama, K. (1995b) *Biochemistry 34*, 3162–3171.

King, R., Jardetzky, O. (1978) Chem. *Phys. Lett. 55*, 15–18.

Kördel, J., Skelton, N.J., Akke, M., Palmer, A.G. & Chazin, W.J. (1992) *Biochemistry 31*, 4856–4866.

Korzhnev, D.M., Orekhov, V.Y. & Arseniev, A.S. (1997) *J. Magn. Reson. 127*, 184–191.

Lefèvre, J.-F., Dayie, K.T., Peng, J.W. & Wagner, G. (1996) *Biochemistry 35*, 2674–2686.

Lipari, G. & Szabo, A. (1982a) *J. Am. Chem. Soc. 104*, 4546–4559.

Lipari, G. & Szabo, A. (1982b) *J. Am. Chem. Soc. 104*, 4559–4570 (1982).

Luginbühl, P., Pervushin, K.V., Iwai, H. & Wüthrich, K. (1997) *Biochemistry 36*, 7305–7312.

Mer, G., Dejaegere, A., Stote, R., Kieffer, B. & Lefèvre, J.-F. (1996) *J. Chem. Phys. 100*, 2667–2674.

Peng, J.W. & Wagner, G. (1992a) *J. Magn. Reson. 98*, 308–332.

Peng, J.W. & Wagner, G. (1992b) *Biochemistry 31*, 8571–8586.

Peng, J.W. & Wagner, G. (1995) *Biochemistry 34*, 16733–16752.

Phan, I.Q.H., Boyd, J. & Campbell, I.D. (1996) *J. Biomol. NMR 8*, 369–378.

Woessner, D.E. (1962a) *J. Chem. Phys. 36*, 1–4.

Woessner, D.E. (1962b) *J. Chem. Phys. 37*, 647–654.

Zheng, Z., Czaplicki, J. & Jardetzky, O. (1995) *Biochemistry 34*, 5212–5223.

10

NMR STUDIES OF PROTEIN SIDECHAIN DYNAMICS

Examples from Antifreeze and Calcium-Regulatory Proteins

Leo Spyracopoulos, Stéphane M. Gagné, Wolfram Gronwald, Lewis E. Kay, and Brian D. Sykes

Department of Biochemistry
MRC Group in Protein Structure and Function
Protein Engineering Network of Centers of Excellence
University of Alberta
Edmonton, Alberta, Canada T6G 2H7

1. INTRODUCTION

NMR spectroscopy is ideally suited for the study of protein dynamics (as well as for other macromolecules). The first reason for this is that NMR is a very high resolution spectroscopic tool which enables the researcher to observe the NMR spectra of all of the individual nuclei in the protein. The second reason is that the characteristic NMR properties of the resonances directly contain dynamic information. Thus multinuclear, multidimensional NMR coupled with the availability of uniformly labeled $^{13}C/^{15}N$ recombinant proteins allows for the determination of dynamic processes at nearly every atomic site in proteins. Information about backbone and sidechain rotational and translational correlation times in proteins may be obtained directly from the measurement of relaxation times of nuclei. The effects of chemical/conformational exchange and reaction kinetics are manifest in NMR lineshapes and provide information about slower timescale dynamics. This is obviously an enormously rich source of detailed information about dynamics.

Dynamics in proteins can occur with amplitudes ranging from 0.01 – 100 Å and timescales from $10^{15} – 10^{-3}$ s^{-1}. These motions include small amplitude fast fluctuations of the atoms, and sidechains; intermediate larger amplitude, slower motions involving helices, domains, and subunits; large amplitude slow motions such as helix-coil transitions, and protein folding. Slower timescale motions can often be described in terms of conformational equilibria between various conformations which includes interactions with ligands or other proteins.

Protein Dynamics, Function, and Design, edited by Jardetzky *et al.*
Plenum Press, New York, 1998

The overall rotational correlation time (τ_R) of the protein is proportional to it's molecular weight (MW). A useful rule of thumb for globular proteins, in water and at 25 °C, is that τ_R (in nanoseconds) is equal to one-half of the MW (in kiloDaltons). Thus:

$$\tau_R = \frac{1}{2} \times MW \qquad (ns)$$

(1)

For example, τ_R is typically 10 nanoseconds (10^{-8} s) for a protein of MW equal to 20,000 Daltons. The correlation time for protein internal motions (τ_{INT}) can range from much faster then τ_R (*i.e.*, τ_{INT} approximately picoseconds to sub nanoseconds) for the internal rotation of amino acid sidechains such as methyl groups all the way up to milliseconds for aromatic sidechains flipping in the most stable proteins. Larger scale conformational changes involving domains or larger sections of proteins generally occur on a longer timescale than sidechain internal motions (*i.e.*, microseconds to seconds). Other processes, such as protein folding or amide proton exchange with solvent occur on even longer timescales of seconds to hours. Generally, the amplitude and nature of the fast timescale internal motions are more well characterized, whereas slower motions are less well understood.

There are several ways to use NMR to measure the dynamics of a protein. These days, most NMR studies of protein dynamics involve relaxation measurements for the heteronuclei (^{13}C and/or ^{15}N) in a labeled protein. This is due to the exceptional resolution of heteronuclear 2D spectra, which are also assigned in the process of structural determination. Longitudinal relaxation given by the rate R_1 ($=1/T_1$), transverse relaxation rate R_2 ($=1/T_2$), and the nuclear Overhauser effect (nOe) mostly reflect the overall and internal motions of the protein on the timescale of the inverse of the resonance frequency of the spectrometer ($1/\omega_o = 10^{-9}$ s for $\omega_o = 500$ MHz). Thus these measurements are sensitive to faster timescales (10^{-11}–10^{-8} s), as indicated in Figure 1.

Micro- to millisecond timescale motions can and do influence transverse relaxation. However, extraction of the contribution of slower rates is more difficult and places greater demands upon the accuracy and precision of the measurements. These motions (10^{-5}–10^1 s) are in the chemical exchange regime and are most easily determined by lineshape measurements or the direct observation of the reaction if the kinetics are slow enough. Overall, this leaves a gap in the timescale range (10^{-8}–10^{-5} s) which is less easily accessible, although methods such as rotating frame relaxation or relaxation dispersion can be useful.

As indicated above, the most common type of study is heteronuclear relaxation for labeled proteins since the heteronuclear spectra are well resolved, the heteronuclei are assigned as part of the structure determination process, and the methodology has been well described in the literature (Farrow *et al.*, 1994). These approaches are described completely in other chapters in this book. Further, these measurements are most often directed towards backbone dynamics (*i.e.*, ^{15}N relaxation).

Methods for the study of sidechain flexibility in the interfacial region of biomolecular recognition processes will be the focus of this chapter. The surfaces of proteins are of paramount importance in determining the nature and affinity of binding of ligands and interaction targets. The majority of the protein surface consists of exposed amino acid sidechains. Furthermore, exposed hydrophobic sidechains are often involved in critical interactions in recognition complexes. In particular, the methyl groups of hydrophobic sidechains are attractive targets for dynamics measurements for the following reasons: they are plentiful, typically well-resolved, offer high sensitivity, and are easily labeled.

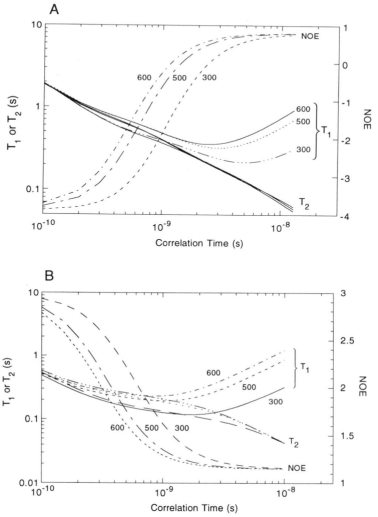

Figure 1. Predicted ^{15}N (A) and ^{13}C (B) relaxation parameters T_1, T_2, and nOe as a function of the overall rotational correlation time τ_R for the protein and for several spectrometer frequencies (300, 500, and 600 Mhz) for backbone amide ^{15}N and backbone ^{13}Cα. Equations from Peng & Wagner (1992) for ^{15}N and Palmer *et al.* for ^{13}C (1991). Relaxation mechanisms include ^{1}H-^{15}N and ^{1}H-^{13}C dipole-dipole as well as ^{15}N and ^{13}C chemical shift anisotropy with parameters from Peng & Wagner (1992) for ^{15}N and Palmer *et al.* for ^{13}C (1991).

Much of the discussion will center around the amino acid threonine, with examples from the calcium regulatory muscle protein troponin-C and the ice-growth regulatory antifreeze proteins. The choice of threonine as a principal example is due to the simplicity of the sidechain geometry, which leads to a straightforward analysis of the dynamics. Furthermore, the sidechain is amphiphillic, consisting of a methyl and hydroxyl group which can be used in intermolecular interactions. In summary, the focus of this article will be on methods to study the role of sidechain flexibility in the interfacial regions of biomolecular recognition complexes. We will discuss two main methodologies—sidechain relaxation

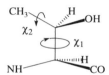

Figure 2. Structure of the sidechain of the amino acid threonine. Note that Thr contains two chiral centers. The vicinal angles that describe the sidechain orientation are indicated.

measurements to determine internal segmental correlation times, and the influence of chemical exchange on the measurement of NMR vicinal coupling constants used to determine sidechain vicinal angles.

2. MODELING THREONINE SIDECHAIN DYNAMICS

When dealing with sidechain internal motions we need to think about the angles and bonds around which motion is possible. The threonine sidechain is simple enough that the internal motion can be modeled explicitly. For more complex amino acid sidechains, it has become customary to use the 'model free' approach as described by Lipari and Szabo (1982a,b). The sidechain for Thr is shown in Figure 2.

Internal motion is possible about the χ^1 and χ^2 vicinal angles in the Thr sidechain. Rotation about the CH_3 group axis is generally very fast ($\tau_{INT} \sim 10^{-12}$ s), leading to a diminution of the effectiveness of the relaxation. For the relaxation studies described below we will focus on the CH_2D group which can occur when a fractionally deuterated protein is prepared, and can be selected exclusively by the pulse sequence (see below). The relaxation of the H, D, or C nuclei on a CH_3 or CH_2D group is reduced by a factor of $[3 \cos^2(\theta_2) -1)/2]^2$ where θ_2 is the angle between the interaction vector for the relaxation mechanism and the angle of internal rotation (see Figure 3).

As a result of this rapid internal rotation, only rotation around the χ^1 angle will be relevant to determining the orientation of the Thr sidechain (and the functionally important CH_3 and OH groups) in the protein.

Rotation around the χ^1 angle is not generally rapid, and is most often discussed in terms of the Newman projection shown in Figure 5.

The populations of the various rotamers for internal rotation around the $\chi1$ angle and rotation rates for the interconversion between rotamers are determined by the potential energy surface for the Thr sidechain. The populations of the Thr rotamers are shown in Figure 6.

3. RELAXATION METHODS FOR STUDYING METHYL SIDECHAIN DYNAMICS

Several research groups have highlighted the importance of hydrophobic amino acid sidechains in the interfacial regions of protein-ligand and protein-protein complexes, and

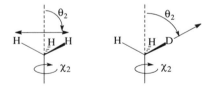

Figure 3. Definition of angles for internal motions. Double arrow indicates proton-proton dipolar interaction vector, single arrow indicates deuterium quadrupolar vector. $\theta_2 = 90°$ for ^1H-^1H dipole-dipole relaxation and $\theta_2 = 70.5°$ for ^2H quadrupolar relaxation.

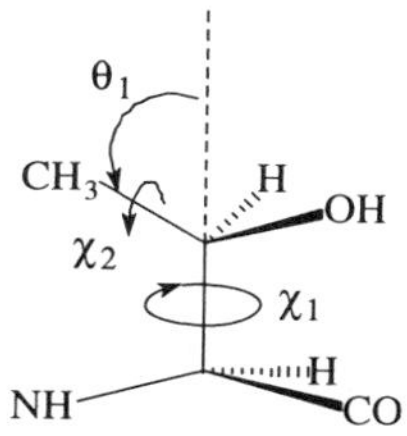

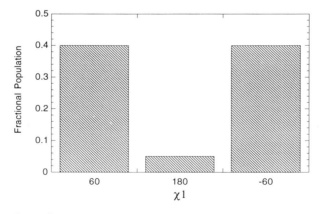

Figure 4. Definition of angle θ_1 for rotation around $\chi 1$.

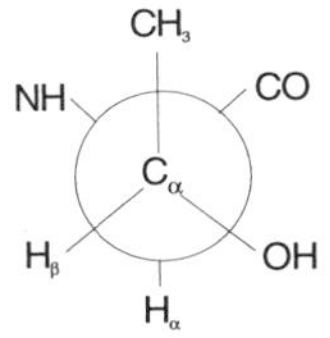

Figure 5. Possible χ^1 rotamer conformations for the Thr sidechain shown as Newman projections. Rotamer shown is for $\chi^1 = 180°$.

attempted to study the role of internal flexibility in determining the affinity and specificity of these interactions. The methyl group plays an important role in these interactions because it is present in the sidechains of Ala, Thr, Leu, Ile, and Val, amino acids whose occurrence is prevalent in proteins. Further, the CH_3 region of a 2D ^1H-^{13}C HSQC NMR spectrum offers the best resolution, highest sensitivity (3 protons) and narrowest linewidths (due to rapid internal CH_3 rotation).

As pointed out by several research groups, however, interpretation of the ^1H and ^{13}C relaxation properties of CH_3 groups is not straightforward because of the existence of cross-correlation effects between dipolar interactions which make the relaxation complex. These effects increase in severity for larger molecular weight proteins and with more pronounced anisotropy of molecular tumbling. Recently, Lewis Kay and co-workers have

Figure 6. Populations for the three most probable rotamer conformations in threonine.

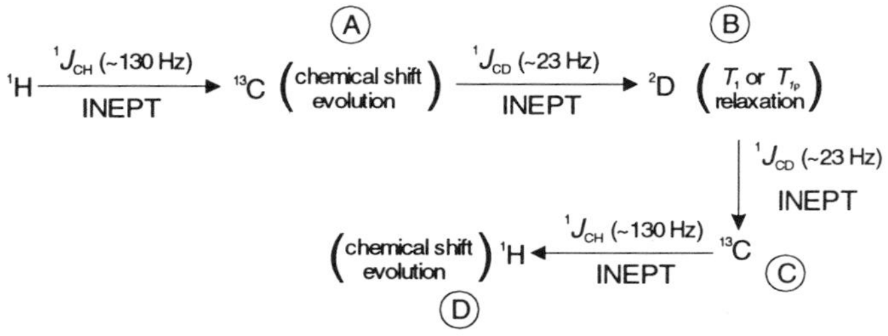

Figure 7. Magnetization transfer pathway for pulse sequences designed for the measurement of deuteron T_1 and $T_{1\rho}$ relaxation.

published an elegant method for studying methyl sidechain dynamics in random fractionally ^{2}H-labeled and uniformly ^{13}C-labeled proteins (Muhandiram *et al.*, 1995). These methods preserve the resolution and sensitivity advantages of the 2D ^{1}H-^{13}C HSQC NMR spectrum mentioned above, but extract the dynamic information by focusing on the relaxation of the ^{2}H nuclei. The relaxation of ^{2}H is dominated by the quadrupolar interaction and therefore reflects the internal rotational motions in a straightforward fashion.

Pulse sequences developed for the measurement of deuterium T_1 and $T_{1\rho}$ (transverse relaxation time in the rotating frame of the spinlock field) are based upon constant time 2D ^{1}H-^{13}C HSQC schemes (Muhandiram *et al.*, 1995). The fate of the methyl group magnetization during the T_1 and $T_{1\rho}$ pulse sequences is outlined in Figure 7.

At point A in the pulse scheme, methyl group magnetization of the form $2(I_{1z}+I_{2z})C_y$ for the CH$_2$D groups is selected exclusively from C_x magnetization arising from CHD$_2$ groups and $4(I_{1z}I_{2z}+I_{1z}I_{3z}+I_{2z}I_{3z})C_x$ magnetization arising from CH$_3$ groups. At point B in the pulse sequence, magnetization from the CH$_2$D groups is transformed into $(I_{1z}+I_{2z})C_zD_z$ in the case of the T_1 experiment or $(I_{1z}+I_{2z})C_zD_y$ for the $T_{1\rho}$ experiment. Although we are interested in measuring the relaxation properties of D$_z$ and D$_y$, the additional delays required to refocus the three spin terms $I_zC_z(D_z$ or D$_y)$ (where I_z= $I_{1z}+I_{2z}$) to D$_z$ or D$_y$ would lead to unacceptable sensitivity losses due to the efficient transverse relaxation of deuterium. The deuterium relaxation properties of the CHD$_2$ groups could also be measured. However, at point B in the pulse sequence, magnetization arising from these groups would involve spin products of two deuterons and more complex expressions are required for a description of their relaxation. In contrast, the CH$_2$D groups will only have magnetization proportional to $I_zC_z D_z$ (or $I_zC_z D_y$) and their relaxation can be described in a simpler fashion.

The decay rates of the three spin terms $I_zC_zD_z$ and $I_zC_zD_y$ are extracted from peak intensities in 2D ^{1}H-^{13}C HSQC spectra recorded as a function of the relaxation delay (point B in Figure 7). The methyl deuteron quadrupole coupling constant is ~165 kHz, significantly greater than the ^{1}H-^{13}C or ^{2}H-^{13}C dipolar interactions and thus dominates the relaxation of the methyl deuteron. Interference between the quadrupolar interaction and the various dipolar interactions (^{1}H-^{13}C and ^{2}H-^{13}C) lead to shifts in the Zeeman energy levels of the deuteron, giving rise to differential relaxation of the different components of the deuteron multiplet. However, the relaxation rates of the different multiplet components are proportional to the square of the sum of interaction coupling constants and we find that the squared ratio of the deuterium quadrupolar interaction to the ^{1}H-^{13}C dipolar interaction is approximately 50, and that of the deuterium quadrupolar interaction to the ^{2}H-^{13}C dipolar interaction is about 135, thus interference effects may be safely neglected (Yang & Kay, 1996).

Additionally, the relaxation of the three spin terms will have contributions from ^1H-^1H dipolar interactions. These contributions are easily accounted for by measuring the decay rate of I_zC_z for the CH_2D groups and subtracting this rate from the decay rate of $I_zC_zD_z$ and $I_zC_zD_y$. Thus, the decay rates of longitudinal (D_z) and transverse (D_y) ^2H relaxation are given by:

$$R_1(D) = \frac{1}{T_1(D)} = R_1(I_zC_zD_z) - R_1(I_zC_z)$$

(2)

$$R_{1\rho}(D) = \frac{1}{T_{1\rho}(D)} = R_{1\rho}(I_zC_zD_y) - R_1(I_zC_z)$$

(3)

In equations [2] and [3], the longitudinal and transverse relaxation rates of the methyl deuteron are given by (Abragam, 1961):

$$R_1(D) = \frac{3}{16}\left(\frac{e^2qQ}{\hbar}\right)^2\left[J(\omega_D) + 4J(2\omega_D)\right]$$

(4)

$$R_{1\rho}(D) = \frac{1}{32}\left(\frac{e^2qQ}{\hbar}\right)^2\left[9J(0) + 15J(\omega_D) + 6J(2\omega_D)\right]$$

(5)

In equations [4] and [5], $e^2qQ/\hbar$ is the quadrupole coupling constant, and $J(\omega)$ is the spectral density at angular frequency ω. The methyl group dynamics can be modeled effectively using the Lipari-Szabo spectral density:

$$J(\omega) = \frac{2}{5}\left[\frac{S^2\tau_R}{1+(\omega\tau_R)^2} + \frac{(1-S^2)\tau}{1+(\omega\tau)^2}\right]$$

(6)

where τ_R is the overall correlation time for molecular tumbling, $1/\tau = 1/\tau_R + 1/\tau_{INT}$, τ_{INT} is the correlation time for internal motions. The order parameter S is expressed as a product of a factor describing rapid methyl rotation and an order parameter describing motion relative to the rotation axis (the rotation axis is collinear with the bond for which $\chi2$ is defined, see Figure 3):

$$S = S_{axis}\left[(3\cos^2\theta_1 - 1)/2\right]$$

(7)

For rapid methyl group rotation and an angle of 70.5° between the C-D bond and the averaging axis, we have $[3\cos^2(\theta_2) - 1)/2]^2 = 0.111$. Thus additional motion about the averaging axis will be reflected in the value of S_{axis}. For $S_{axis}^2 = 1$, the relaxation is due solely to methyl group rotation about the averaging axis and the isotropic tumbling of the protein. For $S_{axis}^2 < 1$, motion in addition to methyl rotation and molecular tumbling is present, and the amplitude of the motion is indicated by the value of S_{axis}^2. A value of $S_{axis}^2 = 0$ for example, would indicate that the motion of the C-D bond is completely unrestricted with respect to the molecular tumbling. Figure 8 shows predicted relaxation times for T_1 and $T_{1\rho}$ at two magnetic fields.

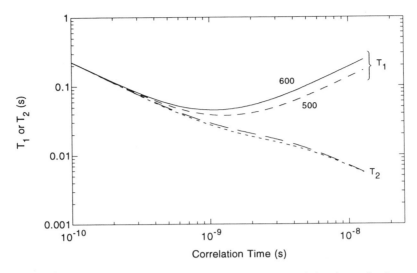

Figure 8. Predicted ^{2}H T_1 and $T_{1\rho}$ as a function of the overall rotational correlation time τ_R for the protein at proton frequencies of 500 and 600 MHz for CH$_2$D methyl groups. Calculations were made using equations [4]-[7] with $\tau_{INT} = 0$, $S_{axis}^2 = 1$, and $\theta_1 = 70.5°$.

4. APPLICATIONS OF DEUTERIUM RELAXATION METHODS TO THREONINE RESIDUES OF THE MUSCLE REGULATORY PROTEIN TROPONIN C

Muscle contraction is initiated by the release of Ca^{2+} ions from the sarcoplasmic reticulum triggering a cascade of events involving several protein structural changes and altered protein-protein interactions (Zot & Potter, 1987). In vertebrate muscle, the conformational change in Troponin C (TnC) resulting from Ca^{2+} binding is the first event in contraction (Gagné *et al.*, 1995). This response is then passed to other components of the thin filament, and ultimately leads to muscle contraction. TnC is a small acidic protein (18 kDa) consisting of two similar globular domains, each containing two Ca^{2+} binding sites. In skeletal muscle, the N-terminal domain of TnC (NTnC; residues 1–90) contains calcium-binding sites I and II and carries out the regulatory function. The calcium-induced structural transition has been previously described (Gagné *et al.*, 1995), and the structural aspect of the mechanism has also been explored (Gagné *et al.*, 1997). An additional aspect required in the understanding of the functional mechanism of TnC is the dynamics of this protein. We present here a portion of this study, the dynamics of the Threonine side-chains in the apo form of NTnC.

The ^{2}H relaxation methods described above have been applied to study sidechain dynamics of NTnC. The sample was [U-^{13}C] and [U-40%-^{2}H] labeled; the deuteration level being chosen to be optimal for the CH$_2$D species. Figure 9 shows a two-dimensional spectrum recorded at 600 MHz using the I$_z$C$_z$ pulse sequence (Muhandiram *et al.*, 1995).

There are five Threonine in NTnC: T4, at the N-terminal end, is mostly unstructured based on structural data (Gagné *et al.*, 1995); T39 is located at the junction of the first β-sheet and the third helix (helix-B); T44 is located in the middle of helix-B; T54 is located at the beginning of helix-C; T72 is located in the second β-sheet. Due to partial overlap, no ^{2}H relaxation data was obtained for T72. We measured the decay rates of I$_z$C$_z$, I$_z$C$_z$D$_z$,

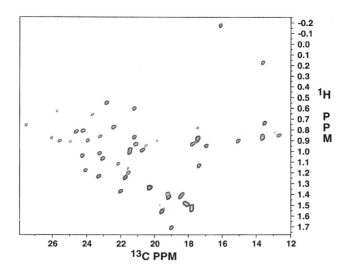

Figure 9. Portion of the 2D contour plot of the first point of the I_zC_z experiment of NTnC recorded at 600 MHz. A small amount of sample degradation has occurred giving rise to a number of extra weak peaks.

and $I_zC_zD_y$ of the CH_2D groups in NTnC, and the decay curves for the four resolved Threonine are shown in Figure 10.

In addition to the ^2H relaxation data, we also recorded ^{15}N relaxation data and determined the overall rotational correlation time (τ_R) to be 4.9 ns. Following this, the backbone S^2 (from the ^{15}N relaxation data) and the methyl S_{axis}^2 were determined for the Threonine residues using the method described above. We have also compiled the solvent accessible surface area of Thr sidechain. This data is summarized in Table 1.

As noted above, T4 is known to be relatively unstructured from the NMR data (Gagné *et al.*, 1995). The flexibility of this residue is also characterized be a small backbone S^2 (0.66). Not surprisingly the S_{axis}^2 of T4 is the smallest of all Threonine at 0.28. Although very small, it does not indicate completely free rotation around χ_1. As illustrated by its sidechain ASA, T39 is the most buried of all five Threonine, and therefore its S_{axis}^2 is relatively high. T44 is part of the same helix as T39 but resides on the exposed side of the helix, and this result in more allowed motion, as illustrated by a smaller S_{axis}^2. The ASA of T54 is the second largest, after T4. Despite the exposition of the methyl group, the S_{axis}^2 of this methyl group is similar to T39, which is mostly buried. This can however be explained by taking a closer look at the structure. The hydroxyl of T54 can potentially form two hydrogen bonds with oxygens located further down the helix (one backbone carbonyl and one Glutamate sidechain). It is therefore likely that the reduced motion we observe for the methyl group, which mainly depends on $\chi1$ restriction, is actually due to hydrogen bonds formed by the hydroxyl group. This hydrogen bond could as an N-terminal helix cap. The position of each threonine in the structure is illustrated in Figure 11.

5. NMR COUPLING CONSTANTS AND LINESHAPE METHODS FOR STUDYING METHYL SIDECHAIN DYNAMICS

From the very earliest days of NMR it has been realized that the three-bond vicinal NMR coupling constants ($^3J_{HH}$) reflect the vicinal bond angle. This relationship between

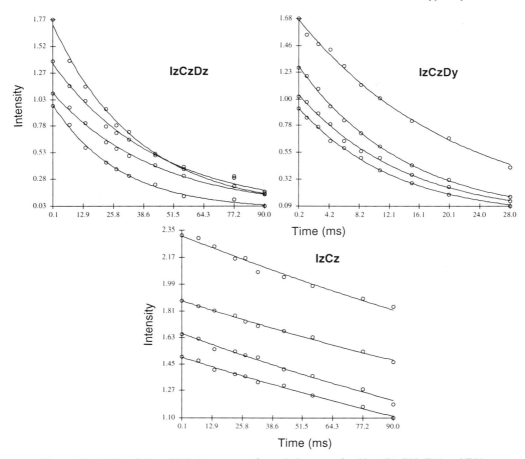

Figure 10. $I_zC_zD_z$, $I_zC_zD_y$ and I_zC_z decay curves for methyl groups of residues T4, T39, T44, and T54.

Table 1. Dynamic characteristics of threonine residues
in the apo N-domain of troponin C [1]

Residue	Backbone S^2 [2]	Side-chain S_{axis}^2 [3]	Side-chain ASA ($\mathring{A}^3$) [4]
T4	0.66	0.28	106.6
T39	0.90	0.69	11.9
T44	0.89	0.60	56.4
T54	0.80	0.69	88.3
T72	0.84	—	37.0

[1] Gagné *et al.*, unpublished data.
[2] Backbone order parameter (S^2) obtained from ^{15}N relaxation data (derived from ^{15}N T_1, T_2, and NOE at 500 and 600 MHz).
[3] Methyl order parameter about the averaging axis (S_{axis}^2) obtained from ^{2}H methyl relaxation data (derived from the decay rates of $I_zC_zD_z$, $I_zC_zD_y$ and I_zC_z of the CH_2D groups). Data for Thr 72 is not available due to partial overlap.
[4] Sidechain solvent accessible surface area (ASA) of Thr residues (PDB accession code 5TNC).

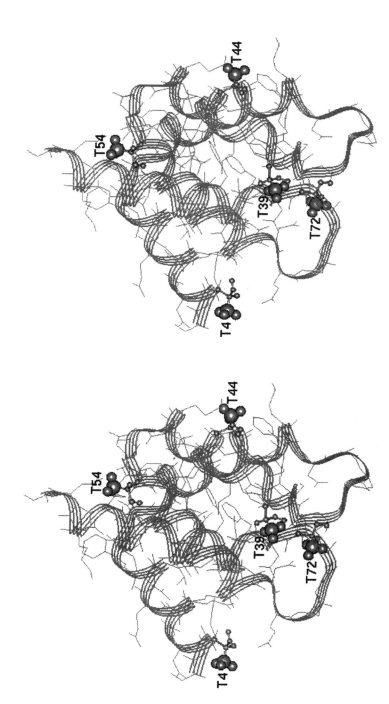

Figure 11. Structure of apo NTnC. The backbone is represented by the ribbon, sidechains by sticks, sidechains of Thr by small ball-and-stick, and the Thr methyl group by larger CPK. Structure from the crystal coordinates (PDB accession code 5TNC).

the value of $^3J_{HH}$ and the angle is generally expressed in terms of a 'Karplus' type equation. In recent years there has been a very rapid increase in the available methods for determination of vicinal angles based upon the measurement of a larger number of coupling constants often involving the heteronuclei in labeled proteins and the resolution of multidimensional, multinuclear NMR spectra. In addition to the classic experiments which involve the direct observation of the splitting in the spectrum (such as DQF-COSY or E. Cosy spectra), new classes of experiments have been developed based upon so called quantitative cross-peak intensities.

In the presence of internal motion, it is most often taken that the vicinal angle in question can only take fixed angles corresponding to the minimum in the potential energy surface of rotation about the bond in question. In this case, the dynamics is reflected in the rate of chemical exchange between the various rotamers. In the case where the exchange is fast $(\tau_{ex} \Delta J \ll 1)$, the observed coupling constants will be a weighted average of the coupling constants for the various rotamers determined by the populations in the various minima.

The $\chi 1$ side chain torsion angles can be obtained by analyzing the pattern of $^3J_{\alpha\beta}$ coupling constants and the relative intensities of the intra residue NOEs involving the αH proton and the two βH^{2,3} protons (Clore & Gronenborn, 1989). In the case of threonines, βH^3 is replaced by the methyl group. The $^3J_{\alpha\beta}$ coupling constants are related to the $\chi 1$ torsion angle *via* the Karplus relationships (Karplus, 1959; Karplus, 1963; Pardi *et al.*, 1984):

$$^3J_{\alpha\beta^2}(\chi 1) = 9.5\cos^2(\chi 1 - 120°) - 1.6\cos(\chi 1 - 120°) + 1.8 \tag{8}$$

$$^3J_{\alpha\beta^3}(\chi 1) = 9.5\cos^2(\chi 1) - 1.6\cos(\chi 1) + 1.8 \tag{9}$$

When both $^3J_{\alpha\beta}$ couplings are small (~3 Hz), then $\chi 1$ must lie in the range 60°. On the other hand, if the $^3J_{\alpha\beta2}$ value is large (~12 Hz) and the other is small, $\chi 1$ can lie either in the region of 180° or -60°. With short mixing time NOESY experiments (Jeener *et al.*, 1979; Macura & Ernst, 1980) that yield stereospecific assignments of the β-methylene protons, it is easy to distinguish between these two possibilities. In the case of threonines, it is only possible to measure $^3J_{\alpha\beta2}$, because βH^3 is the methyl group. Figure 12 outlines the information necessary to determine the $\chi 1$ angles for threonines.

With a precise measurement of the coupling constants in combination with the NOE information it is then possible to get a well defined $\chi 1$ torsional angle (Montelione *et al.*, 1989). However, a sidechain which is not fixed into one conformation, but instead rotates between different rotamers, will show an average coupling constant if the rotation is fast on the NMR time scale (which is valid for side chain dynamics).

$$J_{obs} = \sum_i P_i J_i \tag{10}$$

P_i corresponds to the fractional occupancy of a conformation corresponding to the coupling constant J_i. Based on the expected coupling constants for the different rotamers it is then possible to calculate the occupancy of a specific rotamer from a measured coupling constant and *vice versa*. For a threonine sidechain that has no preference for any of the three rotamers, the following coupling constant would be obtained (equation 11 and Figure 12):

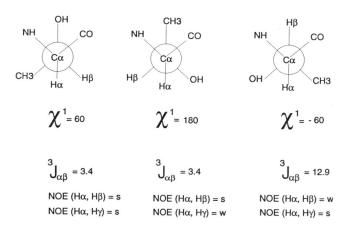

Figure 12. Correlation between $^3J_{\alpha\beta}$ NMR coupling constants and corresponding NOE intensities with $\chi 1$ side chain torsional angles for the three most populated rotamer conformations of threonine. The molecules are shown in the Newman projection with the Cα in the front and the Cβ in the back. For each of the three expected $\chi 1$ values the corresponding coupling constants and NOE intensities are given (s: strong, w: weak). Adapted from Gronwald *et al.*, (1996).

$$^3J_{\text{average}} = \frac{1}{3}(3.4\text{ Hz}) + \frac{1}{3}(3.4\text{ Hz}) + \frac{1}{3}(12.9\text{ Hz}) = 6.6\text{ Hz} \tag{11}$$

In addition to information obtained from coupling constant measurements, the preferred rotameric state of the threonines was also assessed by analyzing the NOE information of the α, β and γ protons in the form of NOE ratios. The NOE ratio is defined according to the following formula:

$$ratio = \frac{(NOE_{H\alpha,H\beta})}{1/3(NOE_{H\alpha,H\gamma})} \tag{12}$$

The Hα, Hγ NOE in the above formula was divided by three to take into account the three threonine methyl protons when compared to the one β proton. As shown in Figure 12, a NOE ratio significantly larger than one corresponds to a $\chi 1$ of 180°, while a NOE ratio significantly less than one corresponds to a $\chi 1$ of -60°. NOE ratios of around one correspond either to a $\chi 1$ of 60° or to a free rotating sidechain. For rotating sidechains the NOE intensities will average similarly to the coupling constants, since the conformation equilibrium is in the fast exchange limit.

$$NOE_{obs} = \sum_i P_i NOE_i \tag{13}$$

Distances calculated from such average NOEs will, of course, be biased towards the smaller distances. For a free rotating sidechain medium intensities would be expected for both the Hα, Hβ and Hα, Hγ NOEs according to Figure 12, which translates into ratios of approximately one.

6. APPLICATION TO FISH ANTIFREEZE PROTEINS

Fish antifreeze proteins (AFPs) are a group of structurally diverse macromolecules that provide freeze-resistance to teleost fish living in ice-laden marine environments (Davies & Hew, 1990). These proteins are present in millimolar concentrations in the serum and are able to depress the freezing point of the blood from -0.8 °C to about -2 °C. The depression of the freezing point is due to the ability of these proteins to inhibit ice crystal growth by binding to the ice surface; a process known as adsorption-inhibition (Raymond & DeVries, 1977). It is generally agreed that the protein binds ice by establishing a hydrogen bonding network with the ice-lattice through a good spatial match of its polar sidechains to the ice surface water molecules. At present, the best characterized and simplest fish antifreeze protein is the α-helical Type I AFP from the winter flounder (isoform HPLC-6) (Davies & Hew, 1990). It is a 37-amino-acid, monomeric α-helical protein. The primary sequence of this protein and other Type I AFP isoforms is built up of a tandemly repeated 11-amino-acid unit. The repeat unit has the consensus sequence TX_2N/DX_7 where X is generally alanine. It appears that the 16.5Å spacing of the $i,i+11$ threonine residues is the critical feature since it matches the 16.7Å distance of the water molecules on the adsorption plane. Since the threonine sidechains form upon ice binding a stable hydrogen bonding network with the ice lattice, their sidechain conformation is paramount to the understanding of the activity of AFPs. Therefore, the flexibility and mobility of the essential threonine sidechains in Type I AFP were investigated near freezing temperature (6 °C) and these observations could be relevant to other systems where surface residues are involved in molecular recognition. A combination of coupling constant measurements obtained from two dimensional DQF-COSY experiments together with NOE data were used to determine the conformational state of the threonine sidechains.

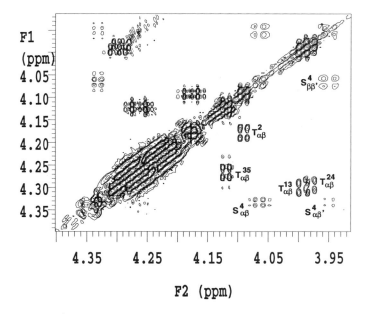

Figure 13. 2D DQF-COSY ¹H NMR spectrum of Type I AFP at 3 °C in D₂O. The Hα-Hβ region containing crosspeaks from T2, S4, T13, T24 and T35 is displayed. Adapted from Gronwald *et al.*, (1996).

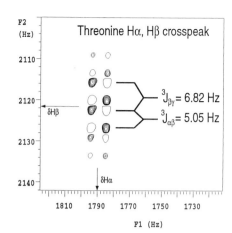

Figure 14. One threonine 2D DQF-COSY Hα-Hβ cross-peak pattern is displayed to show the fine splitting of that cross peak. To distinguish between positive and negative peaks, negative peaks are displayed as open circles.

In all cases, the $^3J_{\alpha\beta}$ coupling constants were determined from the peak separation of well resolved DQF-COSY and/or PE-COSY multiplets. The Hα-Hβ region of a 2D DQF-COSY spectrum of Type I AFP, containing crosspeaks from T2, S4, T13, T24 and T35, is displayed in Figure 13. These are the crosspeaks used for determining the threonine sidechain coupling constants.

Figure 14 gives an example for one well resolved threonine Hα-Hβ crosspeaks of the free amino acid. In the DQF-COSY spectra 512 complex data points were used for a sweep width of 833.3 Hz in both dimensions. To obtain the NOE intensities, the corresponding 2D NOESY cross peaks were integrated using the Varian VNMR processing software.

Based on the $^3J_{\alpha\beta}$ NMR coupling constant values of 7.1, 8.5, 8.5 and 6.8 Hz for residues T2, T13, T24 and T35, respectively, it can be calculated that the regularly spaced ice-binding threonines sample many possible rotameric states prior to ice binding. The lack of a dominant sidechain rotamer is further corroborated by nuclear Overhauser distance measurements for T13 and T24. These data suggest that prior to ice-binding, the threonine sidechains are free to rotate and that a unique pre-formed ice-binding structure in solution is not apparent.

REFERENCES

Abragam, A. (1961) Principles of Nuclear Magnetism, Clarendon Press, Oxford.

Clore, G. M. and Gronenborn, A. M. (1989) Crit. Rev. Biochem. Mol. Biol. **24**, 479–557.

Davies, P.L. and Hew, C.L. (1990) FASEB Journal **4**, 2460–2468.

Farrow, N. A., Muhandiram, R., Singer, A. U., Pascal, S. M., Kay, C.M., Gish, G., Shoelson, S. E., Pawson, T., Forman-Kay, J. D., and Kay, L. E. (1994) Biochemistry **34**, 5984–6003.

Gagné, S. M., Li, M. X., and Sykes, B. D. (1997) Biochemistry, **36**: 4386–4392.

Gagné, S. M., Tsuda, S., Li, M. X., Smillie, L. B., and Sykes B. D. (1995) Nature Struct. Biol. **2**, 784–789.

Gronwald, W., Chao, H., Reddy, D. V., Davies, P. L., Sykes, B. D., and Sönnichsen, F. D. (1996) Biochemistry, **35**, 16698–16704.

Jeener, J., Meier, B. H., Bachmann, P., and Ernst, R. R. (1979) J. Chem. Phys. **71**, 4546–4553.

Karplus, M. (1959) J. Chem. Phys. **10**, 11–15.

Karplus, M. (1963) J. Am. Chem. Soc. **85**, 2870–2871.

Lipari, G., and Szabo, A. (1982a) J. Am. Chem. Soc. **104**, 4546–4559.

Lipari, G., and Szabo, A. (1982b) J. Am. Chem. Soc. **104**, 4559–4570.

Macura, S., and Ernst. R. R. (1980) Mol. Phys. **41**, 95–117.

Montelione, G. T., Winkler, M. E., Rauhenbuehler, P. and Wagner. G. (1989) J. Magn. Reson. **82**, 198–204.

Muhandiram, D. R., Yamazaki, T., Sykes, B. D., and Kay, L. E. (1995) J. Am. Chem. Soc. **117**, 11536–11544.

Palmer, A. G., Rance, M., and Wright P. E. (1991) J. Am. Chem. Soc. **113**, 4371–4380.

Pardi, A., Billeter, M. and Wüthrich, K. (1984) J. Mol. Biol. **180**, 741–751.

Peng, J. W., and Wagner, G. (1992) Biochemistry **31**, 8571–8586.

Raymond, J.A., and DeVries A.L. (1977) Proc. Natn. Acad. Sci. U.S.A. **74**, 2589–2593.

Yang, D., and Kay, L. E. (1996) J. Magn. Reson. **110B**, 213–218.

Zot, A. S., & Potter, J. D. (1987) A. Rev. Biophys. Biophys. Chem. **16**,535–559.

11

MONITORING PROTEIN FOLDING USING TIME-RESOLVED BIOPHYSICAL TECHNIQUES[*]

Kevin W. Plaxco[1] and Christopher M. Dobson[2]

[1]Department of Biochemistry
University of Washington
Seattle, Washington 98195-7350
[2]Oxford Centre for Molecular Sciences
New Chemistry Laboratory
University of Oxford
Oxford OX1 3QT United Kingdom

ABSTRACT

Many of the biophysical techniques developed to characterize native proteins at equilibrium have now been adapted to the structural and thermodynamic characterization of transient species populated during protein folding. Recent advances in these techniques, the use of novel methods of initiating refolding, and a convergence of theoretical and experimental approaches are leading to a detailed understanding of many aspects of the folding process.

1. INTRODUCTION

Native proteins are characterized by a high degree of compactness, an ordered hydrophobic core, a well defined overall architecture and the presence of specific and cooperative interactions between buried side chains. Recent progress in both instrumentation and experimental design has provided unprecedented insights into the evolution of each of these characteristics as an initially disordered and extended polypeptide chain folds to its

---

[*] Adapted with permission from: Plaxco, K.W. and Dobson, C.M. (1996) *Curr. Op. Struct. Biol.* **6**, 630–636.
Abbreviations: ANS, 8-anilino-1-naphthalenesulphonate; CD, circular dichroism; CI2, chymotrypsin inhibitor 2; CIDNP, chemically induced dynamic nuclear polarization; IR, infrared spectroscopy; MS, mass spectrometry; NOE, nuclear Overhauser effect; NMR, nuclear magnetic resonance; 1-D, one-dimensional; 2-D, two-dimensional.

Protein Dynamics, Function, and Design, edited by Jardetzky *et al.*
Plenum Press, New York, 1998

native state. The rapidity and structural heterogeneity of the folding process, however, remains a significant barrier to its experimental characterization. In this chapter we review the techniques applied to the time-resolved experimental characterization of the properties and distribution of partially folded species arising during *in vitro* refolding and the promise that these techniques hold for providing a detailed description of the folding process.

2. THE INITIATION OF FOLDING

Protein folding in the cell follows synthesis of the polypeptide chain on a ribosome. Refolding *in vitro* is more readily initiated by rapidly transferring a protein from denaturing conditions to an environment in which the native conformation is favored. This is often achieved by diluting protein solutions containing denaturant with non-denaturing buffers using a stopped-flow mixing device. Turbulent mixers, such as the Berger ball mixer used in many commercially available instruments, achieve high mixing efficiency by interweaving fine, turbulence generated streams (Berger et al., 1968). The minimum dimensions of these streams have been limited by technical issues, such as cavitation, to a size that denaturants require >100 μs to diffuse from them. Limits on the physical proximity of a detecting cell to the mixer, and the speed with which flow can be stopped without producing shock effects, further increase the dead time of most stopped-flow instruments to >1 ms.

The extremely rapid burst phase events now evident for many proteins are complete within the few millisecond dead time of conventional stopped-flow mixing devices (Eaton et al., 1997). Fortunately, recent technical advances promise significant reductions in initiation dead times. The use of continuous flow devices, which avoid the shock disturbances of high speed stopped-flow, and 'freejet' mixers, which generate small, rapidly diffusing streams by laminar flow through very small orifices, has lowered dead times to tens of microseconds (Bokenkamp et al., 1997; Chan et al., 1997). Non-mixing methods, such as flash photolysis (Jones et al., 1993), optical electron-transfer (Pascher et al., 1996) and temperature jump (T-jump) (Thompson, 1997; Thompson et al., 1997) promise further improvements. Optical electron-transfer, based on the existence of conditions under which the oxidized form of a redox protein is unfolded but the reduced form is native, has been used to initiate the refolding of cytochrome-c in 1 μs by photochemically induced reduction (Pascher et al., 1996). T-jump experiments, based on reversing cold induced denaturation through rapid sample heating, have yielded dead times of 10 μs by electrical discharge heating (Nolting, 1996) and an amazing 20 ns by laser induced heating (Ballew et al., 1996). Applied to the folding of apomyoglobin, laser T-jump has been used to characterize a collapsed state formed in a diffusion limited reaction that is complete in 20 μs (Ballew et al., 1996). When coupled with high speed absorbance, fluorescence and circular dichroism (CD) these new folding initiation techniques will undoubtedly provide important insights into the chemistry of the earliest events in folding.

3. MEASURING COLLAPSE AND CORE PACKING

A general property of protein folding is that an extended and highly disordered polymer chain must collapse to form a compact, globular protein (Miranker and Dobson, 1996). Measures of molecular dimension and core packing (Table 1) are thus critical elements of a complete description of the folding process. Indirect probes of these properties,

Table 1. Many of the biophysical techniques developed to characterize native proteins at equilibrium have now been adapted to the structural and thermodynamic characterization of transient populations during folding. Here we summarize many of the biophysical techniques that have been used in recent years to characterize the folding of a variety of proteins[a]

Physical property	Technique	Approximate resolution	Monitors	Reference
Core packing	Intrinsic fluorescence	µs - ms	The orientation and environment of (predominantly) tryptophan side chains	Engelhard & Evans, 1996
	Ultraviolet absorbance	ms	The orientation and environment of (predominantly) tyrosine side chains	Udgaonkar & Baldwin, 1995
	Extrinsic (ANS) fluorescence	ms	Formation and disruption of organized hydrophobic patches and clefts	Engelhard & Evans, 1995
	Fluorescence quenching	ms	Isolation of tryptophan side chains from hydrophilic fluorescence quenchers	Engelhard & Evans, 1996
	Cystine-quenching	10 s	Protection of cystine side chains from hydrophilic reactants	Ballery et al., 1993
	Kinetic CIDNP	100 ms	Exclusion of specific aromatic side chains from solvent	Hore et al., 1995
Molecular dimensions	Fluorescence anisotropy	ms	Tryptophan side chain mobility and overall molecular dimensions	Jones et al., 1995
	Fluorescence energy transfer	µs - ms	Scalar distance between tryptophan and a covalently attached fluorophore	Rischel & Poulsen, 1995
	Small angle X-ray scattering	<100 ms	The average radius of gyration	Kataoka & Goto, 1996
	Quasi-elastic light scattering	1 s	The average radius of gyration	Feng & Widom, 1994
Secondary structure & persistent hydrogen bonding	Far-UV circular dichroism	ms	Backbone conformation averaged over sequence and population	Evans & Radford, 1994
	Infrared spectroscopy	ns - ms	Backbone conformation averaged over sequence and population	Slayton and Anfinrud, 1997
	Pulse labeling NMR	5-10 ms	Sequence specific formation of stable amide and tryptophan hydrogen bonds	Baldwin, 1993
	Pulse labeling mass spectrometry	5-10 ms	The formation of persistent hydrogen bonds in discrete intermediates	Miranker et al., 1996
Tertiary contacts & native structure	Biological activity	ms - s	The formation of native tertiary structure at the active site	Evans & Radford, 1994
	Interrupted folding	10 ms	The unfolding rate of discrete intermediates as a probe of their stability	Schreiber & Fersht, 1993
	Protein engineering	†	The energetic contributions of side chains to discrete intermediates	Fersht, 1995
	Near-UV circular dichroism	ms	Formation of stable aromatic and disulfide bond tertiary contacts	Evans & Radford, 1994
	Equilibrium 1-D NMR	ms - s	Formation of native tertiary structure at specific residues	Huang & Oas, 1996
	Equilibrium 2-D NMR	ms - s	Formation of native tertiary structure at all resolved amides	Farrow et al., 1996
	Kinetic 1-D NMR	1 s	Formation of specific side chain tertiary contacts	Balbach et al., 1995
	Kinetic 2-D NMR	10 s	Formation of native tertiary structure at all resolved amides	Balbach et al., 1996

[a] A single reference to each method is provided and either reflects a recent review of the subject or an illustrative application of the technique. (†) The time resolution of protein engineering-refolding experiments is limited only by the time resolution of the probe used to monitor the folding of the mutant.

such as changes in the ultraviolet absorbance of aromatic residues (Udgaonkar and Baldwin, 1995), the fluorescence of tryptophan or tyrosine side chains (Engelhard and Evans, 1996) or of extrinsic fluorophores such as 8-anilino-1-naphthalenesulphonate (ANS) (Engelhard and Evans, 1995), have seen widespread application. More direct probes of the exclusion of solvent from the hydrophobic core, by monitoring the accessibility of hydrophilic fluorescence quenchers such as iodide or acrylamide, are also in common use (Engelhard and Evans, 1996). More recently, several site-specific means of monitoring core formation have been developed. These include cysteine quenching (Ballery et al., 1993) and time resolved chemically induced dynamic nuclear polarization (CIDNP) (Hore et al., 1997), which provide an opportunity to monitor the sequestering from solvent of specific, engineered cysteine residues and aromatic residues resolved in 1-D NMR spectra respectively. While these probes represent promising developments, they do not provide an ability to monitor the distribution of individual species in heterogeneous mixtures or to provide a quantitative measure of the dimensions of partially folded conformations. Although little progress has been made on the former problem, several quantitative probes of molecular dimensions are now available.

Time-resolved fluorescence energy transfer, fluorescence anisotropy, small angle X-ray scattering (SAXS) and quasi-elastic light scattering have all been used to provide direct measurement of the dimensions of species arising during folding. The detection of fluorescence energy transfer between a covalently attached fluorophore and a tryptophan side chain, which has been used to attempt direct measurements of the evolution of collapsed species during the refolding of apomyoglobin (Rischel and Poulsen, 1995) and other proteins, is consistent with the hypothesis that these proteins fold via a rapidly formed intermediate of near native compactness. Such studies are, however, limited to proteins that can be modified with suitable fluorophores and only provide measurements of a single scalar distance. Unlike fluorescence energy transfer, the techniques of SAXS (Kawata and Hamagushi, 1991), quasi-elastic light scattering and fluorescence anisotropy provide direct means of monitoring the overall dimensions of macromolecules. SAXS, when implemented with very high flux synchrotron X-ray sources, can provide a measure of the average radius of gyration with 10–100 ms time resolution (Kataoka and Goto, 1996). Recently this has been applied to the refolding of apomyoglobin, again indicating the near native compactness of the major folding intermediate of this protein (Eliezer et al., 1995). Quasi-elastic light scattering, though presently limited to approximately 1 s time resolution, monitors the translational mobility and thus overall dimensions of a macromolecule and has been used to probe the formation of compact states during the refolding of lysozyme (Feng and Widom, 1994). The use of time-resolved fluorescence spectroscopy, not only to monitor molecular dimensions but also to provide a detailed description of the loss of core residue mobility during the refolding of dihydrofolate reductase (Jones et al., 1995), is a recent and very promising approach to the direct, time-resolved detection of molecular dimensions. There appears to be no fundamental reason why these techniques will not prove to be general methods for directly observing the dimensions of a polypeptide chain during protein folding.

4. MONITORING THE FORMATION OF SECONDARY STRUCTURE

Probes of the backbone conformation, such as far-UV CD and pulse labeling hydrogen exchange, have provided a wealth of data on the kinetics of secondary structure formation during folding (Table 1). The recovery of far-UV CD ellipticity is widely considered a definative measure of the average secondary structure content in heterogeneous folding mixtures. More recently time-resolved infrared spectroscopy (IR) has also provided a means of monitoring secondary structure formation kinetics with excellent time resolution (Slayton and Anfinrud, 1997). For many proteins, however, much of the formation of secondary structure occurs in a burst phase during the mixing dead time and thus has not been amenable to direct study. It is fortunate, then, that the invention of extremely rapid methods for the initiation of refolding comes close on the heels of advances in high speed CD (Chen et al., 1997) and IR (Gilmanshin et al., 1997; Slayton and Anfinrud, 1997). Now that the application of high intensity laser light sources to IR spectroscopy and CD spectropolarimetry has produced sub-microsecond time resolution, fundamental questions about the timing of the formation of secondary structure may soon be answered.

While CD provides an estimate of average secondary structure content, it does not provide information on the specific residues involved or the distribution of conformations present. Pulse labeling amide exchange experiments can provide this complementary information by monitoring the formation of stable backbone hydrogen bonds (Baldwin, 1993). Pulse labeling linked to NMR spectroscopy has been used for a number of years as a probe of the sequence specific formation of persistent elements of secondary structure but, like optical methods, the technique cannot resolve individual components from heterogeneous mixtures. Advances in coupling pulse labeling with mass spectrometry (MS) have furthered our understanding of the formation of secondary structure by allowing the observation of resolved molecular species. This has provided a means of characterizing the hydrogen exchange properties of discrete species in heterogeneous populations, as observed, for example, during the refolding of lysozyme (Miranker et al., 1996a). MS, like optical methods, provides data averaged over the entire sequence of a molecule. One approach to localize hydrogen exchange protection monitored by MS is to carry out rapid proteolytic fragmentation prior to mass analysis (Zhang and Smith 1993; Johnson and Walsh, 1994; Smith et al., 1997). Technical advances in MS, however, have proven the feasibility of identifying the sequences of protein cleavage products produced in the gas phase by collision induced dissociation (Anderegg et al., 1994; Miranker et al., 1996b). It may thus soon prove possible to use this approach, in addition to the proteolytic one, to produce sequence specific hydrogen exchange data for discrete species in complex folding populations.

5. DETECTING TERTIARY CONTACTS

Because the formation of ordered regions of native-like structure is thought to be an essential step in protein folding, detecting native tertiary contacts in transient folding populations has been a major goal of folding research. Near-UV CD, which primarily monitors the aromatic side chains immobilized by asymmetric tertiary contacts, has proven an important probe of the recovery of native structure (Evans and Radford, 1994), as, more recently, has IR (Reinstadler et al., 1996). For many proteins, time-resolved assays of the recovery of biological activity (e.g. the binding of fluorescent substrates or inhibitors) can be used to monitor the recovery of a native active site (Udgaonkar and Baldwin, 1995). Interrupted folding experiments, in which transiently refolded mixtures undergo a second unfolding by the rapid addition of denaturant, have been used to detect the formation of material with native stability (Schmid, 1983). This method, which relies on the reasonable assumption that the unfolding rate of a given conformation reflects its thermodynamic stability, has recently been used to monitor the stability of an intermediate in the folding of barnase (Schriber and Fersht, 1993), and to support the existence of parallel pathways in the folding of lysozyme (Kiefhaber, 1995). Protein engineering provides a method of assaying the energetic contributions of specific side chains to the energetics of transient folding intermediates. Major folding intermediates of barnase (Fersht, 1995) and phosphoglycerate kinase (Parker et al., 1996) have been characterized using this technique and it may provide a general probe of the formation of native structural elements.

While protein engineering techniques have provided a means of indirectly monitoring the formation of native tertiary structure at specific sites, a number of NMR techniques are now emerging that provide an ability to monitor simultaneously its formation at multiple sites within a protein. These techniques range from classic line-shape analysis of 1-D data collected from a protein at equilibrium to the direct detection of transient, partially folded species using multi-dimensional NMR to monitor a protein as it refolds. Chemical exchange techniques have been applied to the study of the refolding kinetics of several proteins (Dobson and Evans, 1984; Evans et al., 1989; Roder, 1989; Farrow et al., 1995; Huang and Oas, 1995; Burton et al., 1996). These techniques, based on the effects of the rapid interconversion of different-species on NMR line widths or exchange cross peaks, provide an opportunity to monitor the rate of interconversion of folded and unfolded species on a time scale of milliseconds or less. Analysis of chemical exchange cross-peaks generated in equilibrium mixtures of the folded and unfolded amino-terminal drkSH3 domain has been used to characterized the time constant of folding and to place limits on the extent to which the process is cooperative (Farrow et al., 1995). Analysis of the aromatic region of the 1-D spectrum of a λ-repressor fragment has been used to demonstrate that this all-helix protein folds in a simple, two-state process that is the most rapid refolding characterized to date (Huang and Oas, 1995; Burton et al., 1996).

Line width and exchange methods are limited to conditions under which unfolding and folding are both relatively rapid and under which the folded/unfolded equilibrium is near unity. More recently techniques have been developed to monitor the refolding of proteins by NMR under strongly native conditions. The simplest of these approaches is kinetic 1-D NMR in which refolding is initiated within the magnet and directly monitored by the rapid ($0.5-1$ s^{-1}) acquisition of spectra (Koide et al., 1993; Hoeltzli and Frieden, 1996; Balbach et al., 1995; Roder, 1995). This provides simultaneous information on the rates of formation of all of the tertiary interactions that produce resolved resonances in the spectra (Figure 2). In addition, this strategy can be combined with other NMR techniques, such as the photo-CIDNP methods discussed above. The latter approach, as well as allowing the changes in exposure of residues to solvent to be monitored during folding has also enabled the dead-time in NMR experiments to be reduced to as little as a few tens of milliseconds (Hore et al., 1997). Protein 1-D spectra are, however, often poorly resolved and signal overlap can prove a significant limitation in their applicability. Fortunately, just as multi-dimensional techniques revolutionized the NMR spectroscopy of proteins at equilibrium, there are now promising similar advances in the study of their kinetic processes. One approach is simply to collect series of 2-D spectra if the reaction involved is slow compared to the time needed to accumulate spectra (Baum and Brodsky, 1997). Other strategies are, however, being developed to overcome this constraint. One of these utilizes the fact that the chemical evolution of the species present in a sample during data acquisition will significantly modify intensities and peak shapes in the resulting spectrum. Analysis of the line shapes of individual resonances in heteronuclear 2-D spectra, for example, has allowed simultaneous monitoring of the formation of native structure at individual residues with a time resolution of tens of seconds (Balbach et al., 1996). Such NMR ex-

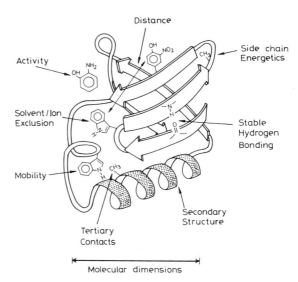

Figure 1. A schematic picture of the characteristics of globular proteins that can be monitored during refolding with time resolution in the second to millisecond range. Other properties that can be monitored but are not indicated include the creation and disruption of organized hydrophobic voids and overall thermodynamic stability. While no individual probe can monitor all of the structural details of a folding intermediate, the use of multiple complementary probes can provide a detailed picture of the distribution of conformations that make up transient folding populations.

periments offer the prospect of studying the folding process at a level of detail much greater than is possible by other time resolved techniques which are limited to at most a few reporter sites. Moreover, in the future one can anticipate the development of methodology based on nuclear Overhauser effects able to provide information in time-resolved mode concerning the detailed structures of the ensemble of molecules present at different stages of a folding reaction (Balbach et al., 1997).

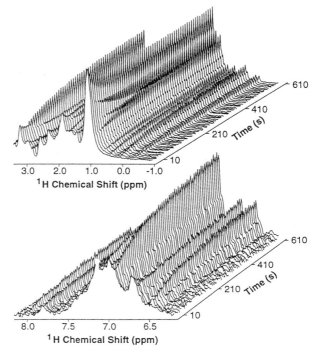

Figure 2. Example of a simple kinetic NMR experiment probing folding. Bovine alpha-lactalbumin dissolved in 6M guanidinium chloride was injected into refolding buffer within an NMR magnet. Spectra were recorded at incremental time points between 1.2 and 620s after the initiation of refolding, and these stack plots show the changes in the spectral regions containing resonances from aliphatic groups (top) and aromatic residues (bottom). The disappearance of the broad and overlapping spectrum of the initially formed species and the well resolved resonance characteristic of the native state are evident. Adapted from Balbach et al., 1995.

6. PROBING THE TRANSITION STATE

A complete description of the folding process requires knowledge of both the structure and energetics of the rate determining conformation. While the ephemeral nature of transition states precludes direct structural studies, the transition state is the conformation of the rate limiting step (or steps) and therefore the kinetics of folding can provide an indirect probe of its structure. The effect of environmental factors and mutations on the kinetics of the recovery of native properties (such as fluorescence) have thus been used to provide a detailed picture of the conformation of this most fleetingly transient species in protein folding.

Environmental factors that affect folding rates have provided valuable clues to the general nature of folding transition states. For example, studies of the temperature dependence of protein folding rates have been used to probe their thermodynamic properties (Oliveberg et al., 1995). Other studies, monitoring the effects of pressure (Vidugiris et al., 1995), denaturants (Chen et al., 1992), ionic strength (Itzaki et al., 1994), deuteration (Parker and Clarke, 1997) and pH (Oliveberg and Fersht, 1996) have been used to define the relative molar volumes and solvent exposed surface areas of the transition states and to probe the contributions of ionizable groups and hydrogen bonding to their energetics. From such studies a general picture is emerging of a typical transition state as a collapsed but relatively poorly packed ensemble of conformations.

Efforts to ascertain the high resolution structure of a folding transition state have focused on protein engineering experiments designed to produce a map of the energetic contributions of specific side chains to the rate limiting step. Comparison of such results with the contributions of these same side chains to the energy of the native state has been carried out in some detail for barnase and chymotrypsin inhibitor 2 (CI2) (Fersht, 1995). These studies, in particular when coupled with theoretical simulations, have provided detailed insights into the structure and heterogeneity of the transition state (Fersht et al., 1994; Daggett et al., 1996) and suggest that small nuclei of native-like structure are involved in the rate determining step of the folding of at least some proteins.

7. FUTURE PROSPECTS

As the number and quality of biophysical techniques with sufficient time resolution increases, there is a concomitant improvement in our detailed knowledge of the folding process. Issues such as cooperativity, collapse and the formation of secondary structure during refolding are becoming well described for a number of proteins. What is still lacking, however, is a means to generate models of the structure and distributions of transient folding populations with good spatial resolution. The next challenge in protein folding lies in how to produce these high resolution models. Several potential approaches now appear feasible.

While no single method can provide a complete picture of the distribution of structures at different stages of a folding reaction (Figure 1), it is clear that multiple complementary approaches must be combined to generate detailed structural models. For example, dynamic light scattering and intrinsic fluorescence can be used to define the average dimensions and degree of core packing, pulse labeling amide exchange can provide information on the location and stability of secondary structure, and NMR and inhibitor binding can be used to define specific tertiary contacts. Such information can thus be combined to develop a detailed picture of the key features of the folding process (Udgaonkar and Baldwin, 1995; Radford and Dobson, 1995; Plaxco et al., 1996; Dobson et al., 1998).

The complementary aspects of theory and experiment in protein folding also suggest a means of providing a high resolution picture of folding pathways. In particular, it is to be hoped that, with the addition of experimentally derived constraints, theoretical simulations (Karplus and Sali, 1995; Onuchic et al., 1995; Shakhnovich et al., 1996) will lead to higher resolution models of the structural transitions taking place during folding with significant predictive value. The use of complementary biophysical approaches, and particularly developments in NMR spectroscopy, to obtain adequate information to constrain theoretical models holds great promise for providing a detailed description and understanding of the folding process.

ACKNOWLEDGMENTS

We wish to thank Jochen Balbach, Julie Forman-Kay, Yuji Goto, Elaine Marzluff, Ken Walsh, Jay Winkler and members of the Dobson group for generously sharing their expertise. This research is supported in part by an International Research Scholars award from the Howard Hughes Medical Institute. The Oxford Centre for Molecular Sciences is supported by the UK Biotechnology and Biological Sciences Research Council, the Medical Research Council and the Engineering and Physical Sciences Research Council.

REFERENCES

Anderegg, R.J., Wagner, D.S., Stevenson, C.L. and Borchardt, R.T. (1994) *J. Am. Soc. Mass. Spec.* **5**, 425–433.
Balbach, J., Forge, V., van Nuland, N.A.J., Winder, S.L., Hore, P.J. and Dobson, C.M. (1995) *Nat. Struc. Biol.* **2**, 865–870.
Balbach, J., Forge, V., Lau, W.S., van Nuland, N.A.J., Brew, K. and Dobson, C.M. (1996) *Science* **274**, 1161–1163.
Balbach, J., Forge, V., Lau, W.S., van Nuland, N.A.J., and Dobson, C.M. (1997) *Proc. Natl. Acad. Sci. USA* **94**, 7182–7185.
Baldwin, R.L. (1993) *Curr. Opin. Struct. Biol.* **3**, 84–91.
Ballew, R.M., Sabelko, J. and Gruebele, M. (1996) *Proc. Natl. Acad. Sci. USA* **93**, 5759–5764.
Ballery, N., Desmadril, M., Minard, P. and Yon, J.M. (1993) *Biochemistry* **33**, 708–714.
Baum, J. and Brodsky, B. (1997) *Folding and Design.* **2**, R53–R60.
Berger, R.L., Backo, B. and Chapman, H.F. (1968) *Review of Scientific Instruments* **39**, 493–498.
Bokenkamp, D., Desai, A., Yang, X., Tai, Y.-C., Marzluff, E.M. and Mayo, S.L. (1997) *J. Anal. Chem.* In press.
Burton, R.E., Huang, G.S., Daugherty, M.A., Fullbright, P.W. and Oas, T.G. (1996) *J. Mol. Biol.* **263**, 311–322.
Chan, C.K., Hu, Y., Takahashi, S., Rousseau, D.L., Eaton, W.A. and Hofrichter, J. (1997) *Proc. Natl. Acad. Sci. USA* **94**, 1779–1784.
Chen, B-L., Baase, W.A., Nicholson, H. and Schellman, J.A. (1992) *Biochemistry* **31**, 1464–1476.
Chen, E.F., Goldbeck, R.A. and Kliger, D.S. (1997) *Ann. Rev. Biophys. Biomolec. Struct.* **26** 327–355.
Daggett, V., Li, A., Itzhaki, L.S., Otzen, D.E. and Fersht, A.R. (1996) *J. Mol. Biol.*, **257**, 430–440.
Dobson, C.M. and Evans, P.A. (1984) *Biochemistry* **23**, 4267–4270.
Dobson, C.M., Sali, A. and Karplus, M. (1998) *Angew. Chem. Int. Ed. Eng.*, In press.
Eaton, W.A., Munoz, V., Thompson, P.A., Chan, C.K. and Hofrichter, J. (1997) *Curr. Op. Struct. Biol.* **7**, 10–14.
Eliezer, D., Jennings, P.A., Wright, P.E., Doniach, S., Hodgson, S.K. and Tsuruta, H. (1995) *Science* **270**, 487–488.
Engelhard, M. and Evans, P.A. (1995) *Protein Sci.* **4**, 1553–1562.
Engelhard, M. and Evans, P.A. (1996) *Folding and Design* **1**, R31-R37.
Evans, P.A., Kautz R.A., Fox, R.O. and Dobson, C.M. (1989) *Biochemistry* **28**, 362–370.
Evans, P.A. and Radford, S.E. (1994) *Curr. Op. Struct. Biol.* **4**, 100–106.
Farrow, N.A., Zhang, O., Forman-Kay, J.D. and Kay, L.E. (1995) *Biochemistry* **34**, 868–878.
Feng, H.-P. and Widom, J. (1994) *Biochemistry* **33**, 13382–13390.
Fersht, A.R., Itzhaki, L.S., elMasry, N.F., Matthews, J.M., Otzen, D.E. (1994) *Proc. Natl. Acad. Sci.*, **91**, 10426–10429.
Fersht, A.R. (1995) *Phil. Trans. R. Soc.* **348**, 11–15.

Freiden, C., Hoettzli, S.D. and Ropson, I.J. (1993) *Prot. Sci.* **2**, 2007–2014.

Gilmanshin, R., Williams, S., Callender, R.H., Woodruff, W.H. and Dyer, B.R. (1997) *Proc. Natl. Acad. Sci. USA* **94**, 3709–3713.

Hore, P.J., Winder, S.L., Roberts, C.H. and Dobson., C.M. (1997) *J. Am. Chem. Soc.* **119**, 5049–5050.

Hoeltzli, S.D. and Frieden, C. (1996) *Biochemistry* **35**, 16843–16851.

Huang, G.S. and Oas, T.G. (1995) *Proc. Natl. Acad. Sci. USA.* **92**, 6878–6882.

Itzhaki, L.S., Evans, P.A., Dobson, C.M. and Radford, S.E. (1994) *Biochemistry* **33**, 5212–5220.

Johnson, R.S. and Walsh, K.A. (1994) *Protein Sci.* **3**, 2411–2418.

Jones, C.M., Henry, E.R., Hu, Y., Chan, C.-K., Luck, S.D., Bhuyan, A., Roder, H., Hofrichter, J. and Eaton, W.A. (1993) *Proc. Natl. Acad. Sci. USA* **90**, 11860–11864.

Jones, B.E., Beechem, J.M. and Matthews, C.R. (1995) *Biochemistry* **34**, 1867–1877.

Karplus, M. and Sali, A. (1995) *Curr. Opin. Struct. Biol.* **4**, 58–73.

Kataoka, M. and Goto, Y. (1996) *Folding and Design* **1**, R107-R114.

Kawata, Y. and Hamaguchi, K. (1991) *Biochemistry* **30**, 4367–4373.

Kiefhaber, T. (1995) *Proc. Natl. Acad. Sci. USA* **92**, 9029–9033.

Koide, S., Dyson, H.J. and Wright, P.E. (1993) *Biochemistry* **32**, 12299–12310.

Miranker, A.P. and Dobson, C.M. (1996) *Curr. Op. Struct. Biol.* **6**, 31–42.

Miranker, A., Robinson, C.V., Radford, S.E. and Dobson, C.M. (1996a) *FASEB J.* **10**, 93–101.

Miranker, A., Kruppa, G.H., Robinson, C.V., Aplin, R.T. and Dobson, C.M. (1996b) *J. Am. Chem. Soc.* **118**, 7402–7403.

Nolting, B. (1996) *Biochem. Biophys. Res. Commun.* **227**, 903–908.

Oliveberg, M., Tan, Y.S., and Fersht, A.R. (1995) *Proc. Natl. Acad. Sci. USA* **92**, 8926–8929.

Oliveberg, M. and Fersht, A.R. (1996) *Biochemistry* **35**, 2726–2737.

Onuchic, J.N., Wolynes, P.G., Luthey-Schulten, Z., Socci, N.D. (1995) *Proc. Natl. Acad. Sci. USA* **92**, 3626–3630.

Parker, M.J., Sessions, R.B., Badcoe, I.G. and Clarke A.R. (1996) *Folding and Design*, **1**, 145–156.

Parker, M.J. and Clarke, A.R. (1997) *Biochemistry* **36**, 5786–5794.

Pascher, T., Chesick, J.P., Winkler, J.R. and Gray, H.B. (1996) *Science*, **271**, 1558–1560.

Plaxco, K.W., Spitzfaden, C., Campbell, I.D. and Dobson, C.M. (1996) *Proc. Natl. Acad. Sci. USA* **93**, 10703–10706.

Radford, S.E. and Dobson, C.M. (1995) *Phil. Trans. R. Soc. Lond. Biol.* **348**, 17–25.

Reinstadler, D., Fabian, H., Backmann, J. and Naumann, D. (1996) *Biochemistry* **35**, 15822–15830.

Rischel, C. and Poulsen, F.M. (1995) *FEBS Lett.* **374**, 105–109.

Roder, H. (1989) *Methods Enzymol.* **176**, 446–473.

Roder, H. (1995) *Nat. Struct. Biol.* **2**, 817–820.

Schmid, F.X. (1983) *Biochemistry* **22**, 4690–4696.

Schreiber, G. and Fersht, A.R. (1993) *Biochemistry* **32** 11195–11203.

Shakhnovich, E., Abkevich, V. and Ptitsyn, O. (1996) *Nature* **379**, 96–98.

Slayton, R.M. and Anfinrud P.A. (1997) *Curr. Op. Struct. Biol.* **7**, 717–721.

Smith, D.L., Deng, Y.Z. and Zhang, Z.Q. (1997) *J. Mass Spectr.* **32**, 135–146.

Thompson, P.A. (1997) In *Techniques in Protein Chemistry VIII.* D.R. Marshak, Ed., Academic Press Inc., San Diego, pp. 735–743.

Thompson, P.A., Eaton, W.A. and Hofrichter, J. (1997) *Biochemistry* **36**, 9200–9210.

Udgaonkar, J.B. and Baldwin, R.L. (1995) *Biochemistry* **34**, 4088–4096.

Vidugiris, G.J., Markley, J.L. and Royer, C.A. (1995) *Biochemistry* **34**, 4909–4912.

Zhang, Z. and Smith, D.L. (1993) *Protein Sci.* **2**, 522–531.

12

NMR STRUCTURE DETERMINATION OF AN ANTIBIOTIC-RNA COMPLEX

Satoko Yoshizawa and Joseph D. Puglisi*

Department of Structural Biology
Stanford University School of Medicine
Stanford, California 94305-5400

1. INTRODUCTION

The role of RNA in the molecular biology of the cell is well established. It has been more difficult to link the functions of RNA to their three-dimensional folds. The ribosome provides an excellent testing ground for hypotheses of RNA structure and function. The functions of the ribosome:, which include reading of the genetic code through the codon-anticodon interaction, catalysis of peptide bond formation, and translocation down the messenger RNA are well characterized and thought to involve RNA. Yet there is little structural information at high resolution on the RNA components of the ribosome.

RNA is the essential component for ribosome function (Noller, 1991; Noller *et al.*, 1992). 16S and 23S ribosomal RNAs contain very highly conserved regions when ribosomes of different organisms are compared (Gutell, 1994), whereas the sequences and numbers of ribosomal proteins vary strongly. The regions of strong conservation within ribosomal RNA often overlap with functional sites identified by biochemical experiments.

The tRNA-mRNA interaction that establishes the sequence of a protein occurs on the small ribosomal subunit. Adjacent mRNA codons interact with the peptidyl tRNA and aminoacyl tRNA in the P and A sites, respectively. The nucleotides within 16S ribosomal RNA that form these sites have been mapped by chemical probing methods (Moazed & Noller, 1990). The A-site bound tRNA protects a limited number of nucleotides in the so-called decoding site (Fig. 1a, b). Two crucial nucleotides, A1492 and A1493, are universally conserved in all ribosomal RNA sequences. The high phylogenetic conservation of this region of 16S ribosomal RNA is consistent with its important function in decoding. In addition, no ribosomal protein has been shown to interact directly with this region of ribosomal RNA (Powers & Noller, 1995).

* Tel. (650) 498-4397; Fax (650) 723-8464; puglisi@stanford.edu

Protein Dynamics, Function, and Design, edited by Jardetzky *et al.*
Plenum Press, New York, 1998

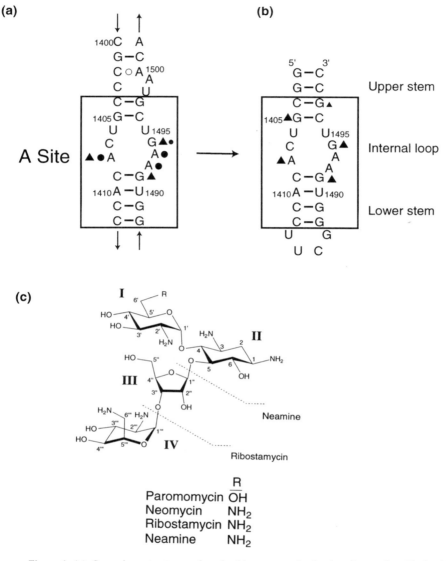

Figure 1. (a). Secondary structure and nucleotide sequence in the decoding region. Nucleotides protected from DMS modification by the addition of A-site tRNA and mRNA are indicated by closed circles. Nucleotides protected by addition of aminoglycoside antibiotics are indicated by triangles. (b) Model oligonucleotide that corresponds to the decoding region A-site. Nucleotides in the oligonucleotide that are protected from reaction with DMS in the presence of 1μM paromomycin are indicated. (c) Chemical structure of the aminoglycoside antibiotic paromomycin. Rings I and II are common elements among all aminoglycoside antibiotics.

We have investigated the structure and function of this region of ribosomal RNA using a combination of NMR structure determination and biochemical methods (Fourmy, 1998; Fourmy, 1998; Fourmy *et al.*, 1996; Recht *et al.*, 1996). The hook that allowed our foray into the ribosome was the aminoglycoside class of antibiotics.

Aminoglycoside antibiotics contain aminosugar and aminocyclitol moieties and bind to ribosomal RNA (Fig. 1c). Aminoglycoside antibiotics cause a decrease in the fidelity of translation and decrease translocation (Davies & Davis, 1968; Davies *et al.*, 1965; Davies *et al.*, 1966) (Edelmann & Gallant, 1977). Aminoglycosides bind to the conserved A-site region of 16S ribosomal RNA. Biochemical experiments, and RNA mutations and modifications that yield aminoglycoside resistance, confirm the direct binding of aminoglycosides to the 16S ribosomal RNA in the small subunit.

2. A MODEL OLIGONUCLEOTIDE FOR NMR STUDIES

To determine the structural origin of aminoglycoside-rRNA recognition, we have used nuclear magnetic resonance (NMR) spectroscopy (Puglisi & Puglisi, 1998). The size limitation for NMR study (about 40 kD) is more than an order of magnitude smaller than the size of the ribosome. The local nature of the aminoglycoside binding site within 16S ribosomal RNA suggested that a small domain of ribosomal RNA would bind to aminoglycoside antibiotics, and be amenable to NMR. Purohit and Stern demonstrated that aminoglycosides bound specifically to a 64nt RNA that spanned the 30S subunit A site (Purohit & Stern, 1994). We have designed a 27nt model oligonucleotide that contains only the ribosomal A site (Recht *et al.*, 1996) (Fig. 1b). A tetraloop sequence was added to the lower stem and two additional G-C pairs were added to the upper stem to facilitate transcription by T7 RNA polymerase.

Before structure determination by NMR, the relevance of a model oligonucleotide to its biological counterpart must be proven. The binding of aminoglycoside antibiotics to the A-site model oligonucleotide and to 30S ribosomal subunits was monitored using chemical probing with the reagent dimethyl sulfate (DMS) (Recht *et al.*, 1996). Upon addition of 1μM paromomycin, positions G1491N7, G1494N7, and A1408N1 are protected from reaction with DMS. The approximate K_d for paromomycin binding is 0.2μM. No footprint on the oligonucleotide was observed with streptomycin, which binds to another region of 16S ribosomal RNA or the polycation spermine , indicating a specific interaction of paromomycin with the oligonucleotide. The set of nucleotides that are protected from reaction with DMS on binding of paromomycin is the same in the 30S subunit and the model oligonucleotide. Changes in the oligonucleotide sequence that are known to disrupt aminoglycoside binding to the ribosome also disrupted high affinity binding to the oligonucleotide. In summary, the biochemical experiments confirmed the specificity of paromomycin binding to the model oligonucleotide, and demonstrated the relevance of the model to ribosomal RNA.

3. RNA NMR SPECTROSCOPY

High resolution NMR structure determination of RNA and its complexes has recently come of age. NMR measurements provide local distance and dihedral angle restraints between pairs of nuclei. Interproton distances are measured using the through-space dipolar coupling, or Nuclear Overhauser Effect (NOE) interaction, which depends on the interproton distance to the inverse sixth power; distances out to 5–6Å can be semi-quantitatively determined. Dihedral angle restraints are determined from the angular dependence of the through-bond J-coupling term. This strategy is exactly analogous to that used for protein structure determination. Unfortunately, the density of proton, and thus restraints, in an RNA is about a third of that for proteins. Therefore, acquisition of a sufficient number of restraints is essential to successful RNA structure determination.

Assignments of RNA resonances are made using isotopically labeled (^{13}C and ^{15}N) RNAs. Biosynthetic methods of preparation of labeled nucleotide triphosphates allows preparation of a labeled NMR sample (Batey *et al.*, 1995). Two distinct classes of protons exist in RNA: exchangeable protons attached to electronegative atoms (N, O) and more numerous nonexchangeable protons attached to carbon. Modern multidimensional NMR methods allow complete proton assignments of RNAs of up to 35 nts. Through-bond magnetization transfer through single nucleotide spin systems and between nucleotides makes these assignments unambiguous (Fig. 2). Labeled RNA also is crucial for studies of RNA-ligand complexes, in which one component is isotopically labeled and the other is not. Isotope-filtered experiments allow selective NOE interactions to be observed only between RNA and ligand. These methodologies were essential to solve the structure of the aminoglycoside-A-site RNA complex.

4. CONFORMATION OF THE FREE A-SITE RNA

The structure of the ligand-free form of the A site oligonucleotide was determined using NMR. Two A-form helices, and the -UUCG- tetraloop form as predicted by the secondary structure. The internal loop is closed by formation of additional base pairs: a U1406-U1495 and C1407-G1494 base pair (Fig. 3). The secondary structure of the RNA is readily determined by examination of the exchangeable imino proton spectrum; one imino proton resonance is expected for each Watson-Crick base pair. For the U-U pair, two imino proton resonances are observed (one from each uracil) and a strong NOE is observed between them, since they are less than 2.5Å apart.

The internal loop is closed by formation of a possible A1408-A1493 base pair. A1492 is bulged in solution. The NMR data indicate that these adenosines do not form a

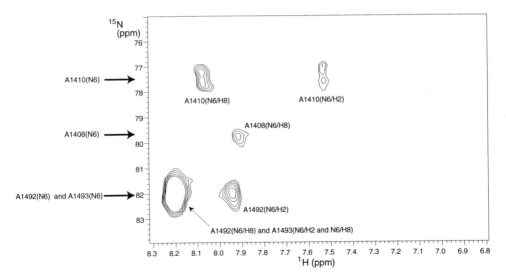

Figure 2. HNC-TOCSY experiment, which correlates amino nitrogen with base protons performed on the paromomycin-RNA oligonucleotide at 15°C. The amino nitrogens of A1408, A1492 and A1493 are indicated with arrows. The A1408(N6-H2) correlation was not observed due to the broadening of the H2 resonance.

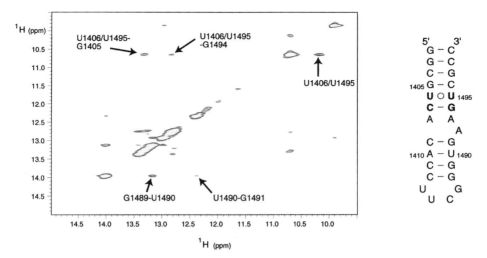

Figure 3. NMR data indicate that the A-form RNA forms a closed internal loop structure. Shown is a NOESY experiment performed in H₂O at 5°C that shows NOEs between exchangeable imino protons. A strong NOE between U1406 and U1495 imino protons indicate formation of a U-U pair, which is followed by a C1407-G1494 pair. The mixing time for this experiment was 250ms.

highly defined structure. There is qualitative evidence for dynamics among these nucleotides, which are stabilized by addition of aminoglycoside antibiotic.

5. STRUCTURE OF THE A-SITE RNA-PAROMOMYCIN COMPLEX

We have studied the structure of a complex of a paromomycin-A-site RNA oligonucleotide complex by NMR. Titration of the A-site RNA with paromomycin indicates a 1:1 stoichiometry of the complex (Fig. 4). As paromomycin is added, a set of imino resonances disappears, and a new set of resonances appears; this indicates that the free and bound forms of the RNA are in slow exchange on the NMR time scale. NMR titrations confirmed the results of biochemical studies. Variants with wild-type affinity for paromomycin behaved in a similar manner as wild-type oligonucleotide. Variants with perturbed affinity for paromomycin give distinctly different behavior.

The structure of the 1:1 complex of paromomycin with the A site oligonucleotide was determined using NMR spectroscopy (Fourmy *et al.*, 1996). A total of 392 NOE derived distance restraints, including 47 intermolecular restraints, were used to calculate the solution structure. The converged structures are well-defined within the aminoglycoside binding site. The core region, which includes G1405 to A1410, U1490 to C1496 and all 4 rings of paromomycin has a root mean squared deviation of 0.61Å among the 20 converged structures. The solution structure thus provides an atomic-level view of how aminoglycosides bind to their ribosomal target (Fig. 5).

Paromomycin binds in the major groove of the A-site RNA, within a binding pocket formed by the asymmetric internal loop. Rings II, III and IV form a linear array along the major groove and ring I fits into a specific pocket formed by an A1408-A1493 base pair and the bulged adenosine A1492. Rings I and II are the best defined portions of paromomycin, and direct specific interaction with the RNA (Fig. 6), consistent with the conservation of

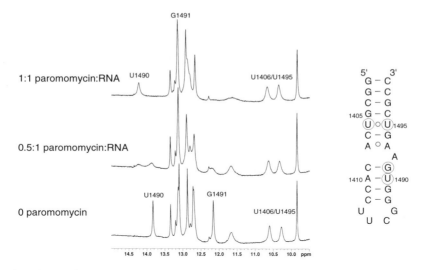

Figure 4. Paromomycin forms a 1:1 complex with the A-site oligonucleotide. The imino proton spectrum for the RNA is shown at 0, 0.5:1 and 1:1 stoichiometry of paromomycin to RNA. A second set of peaks is observed, which indicates that the free and bound forms of the RNA are in slow exchange on the NMR time scale.

these rings in different aminoglycosides that target the ribosomal A site. Rings III and IV vary among aminoglycosides, and these rings are more disordered in the NMR structure.

Comparison of the free and bound forms of the A-site oligonucleotide indicates only a minor conformational change upon paromomycin binding. As in the free form, the internal loop is closed by the U1406-U1495 and C1407-G1494 base pairs. The conformation of A1408, A1492 and A1493 is fixed compared to the free form. A1408 and A1492 form a base pair, with the Watson-Crick face of A1408 interacting with the Hoogsteen face of A1493. The phosphodiester backbone between G1494 and G1491 forms the lip of the drug binding pocket. Binding of paromomycin in the major groove displaces the bases of A1408, A1492 and A1493 towards the minor groove, which may have implications for aminoglycoside-induced miscoding (see below).

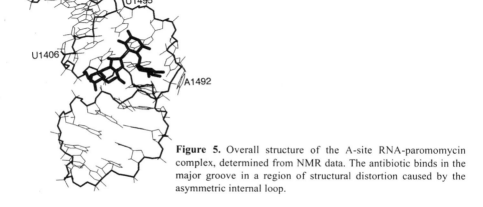

Figure 5. Overall structure of the A-site RNA-paromomycin complex, determined from NMR data. The antibiotic binds in the major groove in a region of structural distortion caused by the asymmetric internal loop.

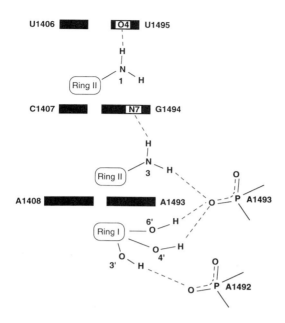

Figure 6. Schematic of intermolecular contacts between paromomycin and the A-site RNA oligonucleotide.

The structure explains activity measurements on aminoglycoside antibiotics. Rings I and II of paromomycin direct specific interaction with the A-site RNA (Fig. 6). Two functionally important amino groups in ring II form hydrogen bonds with the N7 of G1494 and the O4 of U1406. Ring I is located within the binding pocket formed by the A1408-A1493 base pair and A1492. Chemical groups in ring I contact the phosphodiester backbone between G1494 and G1491. No direct hydrogen bonding contacts to RNA bases are made by ring I. The bottom of the binding pocket for ring I is formed by the C1409-G1491 base pair, and ring I is positioned above the base moiety of G1491.

Rings III and IV are less crucial for the antibiotic action of aminoglycosides. Rings III and IV can be deleted or changed and antibiotic action is maintained, albeit at a higher effective concentration. Rings III and IV are located in the major groove of the lower helical stem near base pairs C1409-G1491 and A1410-U1490, but make only electrostatic interactions with the phosphodiester backbone of the RNA helix.

We have studied the interaction of related aminoglycoside antibiotics neomycin, ribostamycin and neamine to the A-site oligonucleotide. Neomycin is similar to paromomycin, except that the 6'-OH is changed to NH_2. NMR and biochemical methods show that neomycin binds to A-site RNA with similar affinity and in a similar manner as paromomycin. Ribostamycin and neamine bind with lower affinity, but the presence of specific drug-RNA NOEs demonstrated that they bind in the same location as paromomycin. In summary, the structural data on diverse aminoglycosides supports the essential role of rings I and II in rRNA interaction.

The high sequence conservation of the A-site RNA creates a difficult target for specific interference of prokaryotic over eukaryotic organisms. Aminoglycoside antibiotics cause miscoding and inhibit translation at 10–50-fold lower concentration for prokaryotic ribosomes than eukaryotic ribosomes. The aminoglycoside binding site is made up of two halves: one contains universally conserved nucleotides and the second contains variable nucleotides. The shape of the aminoglycoside binding pocket changes. However, A1408 is replaced by a G in all eukaryotic sequences. This single mutation reduces the binding affinity of paromomycin to the model oligonucleotide by 15-fold (Recht *et al.*, 1996). The specific geometry of the A1408-A1493 base pair is required to form the binding pocket for ring I and cannot be replaced by a G1408-A1493 pair of similar geometry. In addition, all higher eukaryotic ribosomes contain a mispair at the 1409–1491 position, which forms the bottom of the binding pocket. The combination of the G1408 substitution and 1409–1491 mispair explains the specific action of aminoglycosides on prokaryotic organisms.

6. STRUCTURE EXPLAINS RESISTANCE MECHANISMS

Resistance is widespread for aminoglycosides. Enzymatic modification of aminoglycosides is the most prevalent mechanism of resistance. These enzymes modify chemical groups on rings I and II, and include acetyl-transferases, phospho-transferases and adenyl-transferases (Fig. 7a). Antibiotic-producing organisms must protect themselves from their own antibiotics, and aminoglycoside producing organisms encode methyl transferases that specifically modify ribosomal RNA to yield resistance (Fig. 7b). These enzymes include an A1408 N1 methylase that gives broad spectrum aminoglycoside resistance and G1405 N7 methylase that gives resistance only to gentamicin and kanamycin. Several ribosomal

(a)

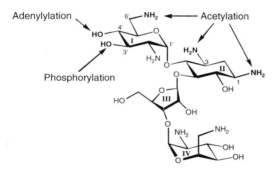

(b)

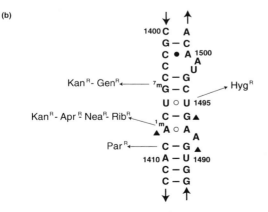

Figure 7. (a) Targets of aminoglycoside modification enzymes on neomycin. (b) Modifications of 16S ribosomal RNA that yield antibiotic resistance. Sites of methylation, and the spectrum of resistance that results are indicated for G1405 and A1408. RNA mutations at 1409–1491 and 1495 that give resistance are also shown.

RNA mutations have been isolated that give aminoglycoside resistance, including the 1409–1491 mispair, and U1495 to C mutation.

The paromomycin-A-site RNA structure explains the origins of these resistance mechanisms. The aminoglycoside chemical groups that are enzymatically modified are in regions of close RNA-aminoglycoside contact, such that acetylation, phosphorylation and adenylylation would disrupt specific hydrogen bonding and electrostatic contacts. RNA modification enzymes also target critical elements for antibiotic binding. Methylation of A1408 N1 disrupts the A1408-A1493 pair that forms the binding pocket for ring I. The N7 of G1405 is distant from paromomycin in the NMR spectrum. However, methylation of G1405 causes resistance only to gentamicin and kanamycin, whose ring III is attached to ring II at position 6. Our unpublished NMR structure determination of gentamicins with the A-site RNA indicate that ring III is in proximity to G1405(N7).

7. IMPLICATIONS FOR RIBOSOME FUNCTION

The miscoding properties of aminoglycosides provide insight into the natural functions of the ribosomal A site. Binding of the aminoglycoside antibiotics decreases the dissociation rate constant for A-site bound tRNAs (Karimi & Ehrenberg, 1994), and the increased affinity of the A site is consistent with the miscoding induced by aminoglycosides. Since ribosome contacts with the A-site tRNA-mRNA complex are an essential feature of models for translational accuracy, aminoglycosides may favor these contacts. In the paromomycin complex, A1492 and A1493 are specifically positioned with their N1s pointing in the minor groove. In the absence of antibiotic, the adenines do not adopt a fixed conformation. The N1 positions of these two universally conserved adenines A1492 and A1493 are protected from reaction with dimethyl sulfate upon binding of A-site tRNA and mRNA (Moazed & Noller, 1990). Based on the conformational change induced on antibiotic binding, and the physical measurements mentioned above, we proposed that the bound conformation represents a high affinity conformation for tRNA binding in the A site (Fourmy et al., 1996). The N1 positions of A1492 and A1493 are positioned such that they can hydrogen bond with two 2'OH groups in the RNA helix formed by the messenger RNA in the codon-anticodon complex.

The proposed mechanism presents specific details for the contribution of the ribosome to the accuracy of protein synthesis. Contacts between the ribosome and the codon-anticodon complex must occur in a sequence independent manner. All cognate codon-anticodon complexes form at least two Watson-Crick pairs in an A-form geometry, and thus should present two 2'OH in a similar orientation. A non-cognate complex would not present the 2' -OH in the proper orientation. In the absence of aminoglycoside antibiotic, the contacts between A1492/1493 and the two 2'OH would drive the conformational change of the A site rRNA. The small free energy cost of the conformational change is overcome by the binding of an aminoglycoside antibiotic in the A site. The proposed contacts are consistent with limited biochemical data. Changes of universally conserved A1492 and 1493 yield lethal phenotypes and substitutions of 2'OH with deoxy nucleotides in the A site codon decrease the binding affinity of the cognate tRNA.

8. CONCLUSIONS

The structure of the paromomycin-A-site RNA complex explains how aminoglycoside antibiotics bind to the ribosome. The structure of the RNA-paromomycin complex provides a point of departure for rational drug design. What remains unanswered is how the aminoglycosides kill cells and how the ribosome insures high fidelity. Our future experiments intend to address these points.

ACKNOWLEDGMENTS

We would like to thank members of the Puglisi group who have participated in this project and Prof. Harry Noller and members of his research group at UC Santa Cruz for discussion of ribosome structure and function. Supported by grants from the Packard Foundation, Deafness Research Foundation, National Institute of Health (GM51266-01A1), Lucille P. Markey Charitable Trust, and Japan Society for the Promotion of Science.

REFERENCES

Batey, R. T., Battiste, J. L. & Williamson, J. R. (1995). Preparation of isotopically enriched RNAs for heteronuclear NMR. *Methods Enzymology.* **261**, 300–323.

Davies, J. & Davis, B. D. (1968). Misreading of ribonucleic acid code words induced by aminoglycoside antibiotics. *J. Biol. Chem.* **243**, 3312–3316.

Davies, J., Gorini, L. & Davis, B. D. (1965). Misreading of RNA codewords induced by aminoglycoside antibiotics. *Mol. Pharmacol.* **1**, 93–106.

Davies, J., Jones, D. S. & Khorana, H. G. (1966). A further study of misreading of codons induced by streptomycin and neomycin using ribopolynucleotides containing two nucleotides in alternating sequence as templates. *J. Mol. Biol.* **18**, 48–57.

Edelmann, P. & Gallant, J. (1977). Mistranslation in *E. coli. Cell.* **10**, 131–137.

Fourmy, D., Recht, M. I., Puglisi, J. D. (1998). Binding of neomycin-class aminoglycoside antibiotics to the A Site of 16S rRNA. *J. Mol. Biol.* **277**, 347–362.

Fourmy, D., Yoshizawa S., Puglisi, J. D. (1998). Paromomycin binding induces a local conformational change in the A site of 16S rRNA. *J. Mol. Biol.* **277**, 333–345.

Fourmy, D., Recht, M. I., Blanchard, S. C. & Puglisi, J. D. (1996). Structure of the A site of *E. coli* 16S rRNA complexed with an aminoglycoside antibiotic. *Science.* **274**, 1367–1371.

Gutell, R. R. (1994). Collection of small subunit (16S- and 16S-like) ribosomal RNA structures: 1994. *Nucleic Acids Res.* **22**, 3502–3507.

Karimi, R. & Ehrenberg, M. (1994). Dissociation rate of cognate peptidyl-tRNA from the A-site of hyper-accurate and error-prone ribosomes. *Eur. J. Biochem.* **226**, 355–360.

Moazed, D. & Noller, H. F. (1990). Binding of tRNA to the ribosomal A and P sites protects two distinct sets of nucleotides in 16 S rRNA. *J. Mol. Biol.* **211**, 135–145.

Noller, H. F. (1991). Ribosomal RNA and translation. *Ann. Rev. Biochem.* **60**, 191–227.

Noller, H. F., Hoffarth, V. & Zimniak, L. (1992). Unusual Resistance of Peptidyl Transferase to Protein Extraction Procedures. *Science.* **256**, 1416–1419.

Powers, T. & Noller, H. F. (1995). Hydroxyl radical footprinting of ribosomal proteins on 16S rRNA. *RNA.* **1**, 194–209.

Puglisi, E. V. & Puglisi, J. D. (1998). Nuclear magnetic resonance spectroscopy of RNA. In *RNA Structure and Function*, (R. W. Simons and M. Grunberg-Manago, eds), pp. 117–146, Cold Spring Harbor Press.

Purohit, P. & Stern, S. (1994). Interactions of a small RNA with antibiotic and RNA ligands of the 30S subunit. *Nature.* **370**, 659–662.

Recht, M. I., Fourmy, D., Blanchard, S. C., Dahlquist, K. D. & Puglisi, J. D. (1996). RNA sequence determinants for aminoglycoside binding to an A-site rRNA model oligonucleotide. *J. Mol. Biol.* **262**, 421–436.

COURSE ABSTRACTS

DEFINING THE CALCIUM-BINDING SITES IN THE LIGAND-BINDING DOMAIN OF THE LDL RECEPTOR

A. Atkins[1], I. Brereton[2], P. Kroon[1] and R. Smith[1]

[1]Biochemistry Department and [2]Centre for Magnetic Resonance, University of Queensland, 4076, Australia

The LDL receptor mediates the uptake of cholesterol through calcium-dependent interactions with the lipoproteins apo E and apo B-100. The ligand-binding domain of the receptor is comprised of seven imperfect repeats of ~40 amino acids, each of which contains six conserved cysteines involved in intra-domain disulfide bonds, in addition to conserved acidic and hydrophobic residues. ^{1}H NMR conformational studies on the first[1] and second[2] repeat identified a similar fold involving a β-hairpin structure followed by a series of turns.

The conformations are strongly dependent on the presence of calcium; despite the three intramolecular disulfide bonds, the NMR amide resonances in the absence of calcium are poorly dispersed. We have investigated the role of calcium in determining the native conformation of the first repeat (LB1). It was found that calcium was required for the formation of the native disulfide bonds. In the presence of calcium, the repeat folds rapidly and exclusively into the native conformation, as determined by HPLC. In contrast, in the absence of calcium, equilibrium is only reached after several days, and multiple, fully oxidised conformers having non-native disulfide bonds are present. These findings are in agreement with studies on the fifth repeat[3], where calcium-dependent folding was also observed.

The calcium-binding site is currently unknown, however, sequence conservation, calcium-sensitive downfield-shifted resonances, and the presence of unexplained slow-exchange amide protons implicate acidic residues in the C-terminal half of the peptide in the coordination of the metal ion. Investigation of the pH titration behaviour of the β and γ protons of Asp and Glu residues, respectively, further delineated the binding site.

We have endeavoured to more precisely define the ion-binding site by replacing the calcium with either paramagnetic or NMR-observable ions previously found to successfully mimic calcium. The calcium in LB1 is in slow exchange on the NMR time scale;

therefore, to employ paramagnetic ions to map the binding site, they must bind at the same site with high affinity. The addition of 0.05 mole equivalent of Yb3+, in the absence of calcium, selectively broadens specific residues in the peptide, indicative of the fast exchange regime. Similarly, Mn2+ is in fast exchange, with small amounts of the ion broadening all resonances apparently nonspecifically, while Gd3+ defined a low-affinity (fast exchange) metal-binding site in the N-terminal half of the sequence. Cd2+, which has been used extensively as a calcium substitute, did not bind to the calcium-binding site on the peptide, as judged by the comparison of NMR spectra.

These findings are consistent with a calcium-binding site in the C-terminal half of the peptide and implicate specific acidic residues, in agreement with the proposed site in the fifth repeat[3].

REFERENCES

1. N. L. Daly, M. J. Scanlon, J. T. Djordjevic, P. A. Kroon and R. Smith (1995) *PNAS 92*, 6334–6338.
2. N. L. Daly, J. T. Djordjevic, P. A. Kroon and R. Smith (1995) *Biochem. 34*, 14474–14481.
3. S. C. Blacklow and P. S. Kim (1996) *Nature Structural Biology 3*, 758–761.

THE WATER STRUCTURALISATION IN BIOLOGICAL TISSUES STUDIED BY PULSE NMR

Daniela Baciu[1] and Ioan Baicu[2]

[1]Department of Biophysics, Apollonia University, and [2]Department of Biophysics, University of Medicine and Pharmacy "Gr. T. Popa", Iasi, Romania

Relaxation times, experimentally determined, demonstrate the presence of many phases of structuralisation of water. There are a minimum of three ways in which water is bound with protein in the tissues. The first type of binding with a short relaxation time has the strongest bond with the proteins compared to the second or the third types. Binding where there is the greatest distance involves a small quantity of water and the water is less structured. This water structuralisation depends on the type of tissue, external factors, stress, cancer, pathological or physiological changes, physics factors, ultrasounds, x-rays, etc. Where the water is bound with the smallest changes, the bonds are strongest and the relaxation times are the shortest. Our experiments demonstrated two types of variations in water binding:

In the first case there were only quantitative changes—the relaxation time stayed the same, but different amounts of water were bound. The water-protein systems were constant, but different quantities of protein versus water were bonded.

In the second case the relaxation times were changed. This demonstrates that the protein design and the water structure changed.

These two types of changes in the bonding of water—functional changes (reversible, as in the first case) and pathological changes (irreversible, as in the second case)—can be seen in PMR tomography.

MOLECULAR MODELING OF AMPHOTERICIN B TRANS-MEMBRANE CHANNEL

M. Baginski,[a,b] H. Resat,[a,c] and J. A. McCammon[a]

[a]Department of Chemistry and Biochemistry, and Department of Pharmacology, University of California at San Diego, La Jolla, CA 92093-0365, USA; [b]Technical University of Gdansk, Narutowicza St 11/12, 80–952 Gdansk, Poland; [c]Department of Physics, Koc University, Istinye, Istanbul 80860, Turkey

Amphotericin B (AmB) is a powerful, clinical, anti-fungal antibiotic used to treat systemic infections. It forms ionic trans-membrane channels in fungal cells. The channel is formed from several antibiotic and sterol molecules. These antibiotic-sterol channels are responsible for the leakage of ions which causes cell death. The detailed molecular properties, structure, and formation of amphotericin B channels are still unknown. In the present study, two molecular dynamic simulations of a particular model of AmB-cholesterol channel were performed. The water and phospholipid environment were explicitly included in our simulation. It was found that mainly the intermolecular hydrogen bonds (especially those between amino and carboxyl groups) keep the channel stable in its open state. Our study also revealed the important role of the intermolecular interactions between the hydroxyl groups of the channel forming molecules. Particularly, it was found that some hydroxyl groups may be new "hot spots" potentially useful for chemotherapeutic investigations. Our results were also discussed with regard to some available experimental data (biochemical and NMR). Our data helps us to understand why some certain antibiotic derivatives are less active and the role of the membrane lipids and cholesterol in the channel stability.

RELATION BETWEEN PROSTHETIC GROUP CONFIGURATION AND PROTEIN CONFORMATION: A HOLE BURNING STUDY

E. Balog,[1] R. Galantai,[1] K. Kis-Petik,[1] M. Kohler,[2] and J. Friedrich[2]

[1]Institute of Biophysics, Semmelweis University of Medicine, Budapest, Hungary and [2]Technical University, Munich, Germany

In case of the proteins which need prosthetic groups for functioning, it is an interesting problem to elucidate the relation between the prosthetic group configuration and the associated protein conformation. In this work pressure tuning and Stark-effect hole burning experiments were carried out on Mg-mesoporphyrin (Mg-MP) substituted horseradish peroxidase (HRP) to study this relation. Pressure tuning experiments demonstrated the presence of two different MgMP conformers stabilized by the same protein conformation. For both configurations Qx-Qy splitting is observed, which demonstrates the porphyrin ring deformation. In this way the pocket field breaks the effective inversion symmetry of the chromophore by inducing a dipole moment as we measured in Stark-effect experiments. By adding a substrate to HRP, which binds in the heme pocket of the enzyme, only one conformer can be detected (with an increased Qx-Qy splitting).

STRUCTURAL APPROACH TO THE STUDY OF THERMAL STABILITY OF UBIQUITIN

Yael Pazy Benhar,[1] George I. Makhatadze,[2] and Gil Shoham[1]

[1]Department of Inorganic Chemistry and Laboratory for Structural Chemistry and Biology, The Hebrew University of Jerusalem, Jerusalem 91904, Israel and [2]Department of Chemistry and Biochemistry, Texas Tech University, Lubbock, Texas 79409-1061

Ubiquitin (Ub) is a small protein (76 amino acids, 8565 daltons) which plays an important role in many biochemical processes in both animals and plants, while maintaining a large degree of amino acid sequence and structural conservation. It is present in the nuclei, the cytoplasm, and the membrane of the cell, and its main role is in the process of targeting proteins for degradation. Some of the most interesting qualities of ubiquitin are related to its high resistance to extreme thermal and chemical conditions. Ub shows exceptional structural and functional stability, and is capable of rapid renaturation after exposure to high temperatures, chemical denaturants and acidic/basic environments.

In the present study we tried to find the main factors that affect stability loss in Ub unstable mutants, using molecular modeling methods. Such studies, performed during dynamics runs (mutant and native structures), enabled us to demonstrate: (1) A comparative approach that corrects for procedural faults in determining an unknown structure; (2) A procedure that allows the determination of differences in domains flexibility during dynamics calculations; (3) A possible explanation for the loss of thermal stability in a mutated ubiquitin.

Our analysis, of the L71I Ub mutant, implies that the flexibility of the carboxy terminus "arm" (shown to be involved in Ub-substrate interactions) has been lowered in the mutant structure. Also observed was a decrease in the flexibility of the region around Lys48, the residue involved in the formation of the poly-Ub "tree". In contrast, a flexibility increase was observed for two loops closing on the hydrophobic core. Increased flexibility of these loops in the L71I mutant, during heating, could enable solvent penetration into the hydrophobic core, and hence to the collapse of the three dimensional rigidity of the molecule at higher temperatures. Preliminary analysis of other mutants show similar effects in different areas of the protein, depending on the site of the mutation. These and related results correlate well with calorimetric studies and indicate that the rigidity of the hydrophobic core is one of the main stabilizing factors in the exceptional thermostability of ubiquitin.

STRUCTURE DETERMINATION OF PLASTOCYANIN WITH NMR

A. Bergkvist and B. G. Karlsson

Institutionen för Biokemi och Biofysik, Chalmers Tekniska Högskola, 413 90 Göteborg, Sweden

Plastocyanin is a link in the electron transport chain in photosynthesis. It is a small (10.5 kD), water soluble protein and is thus suitable for NMR studies. There is currently no well determined structure of spinach plastocyanin, but an assignment of most of the protons has been reported (Driscoll, *Eur. J. Biochem.*, 1987, **170**, 279). Other plastocyanin structures are available, notably one crystal structure of poplar and NMR determined structures of French bean, parsley and *Anabaena variabilis*.

At the Department of Biochemistry and Biophysics, a mutagenesis program has been launched in order to study the structure-function relationship of spinach plastocyanin. As a part of this program we have decided to complete the assignments and determine the structure of wildtype spinach plastocyanin and some mutants. ^{15}N labelling of the protein has enabled us to perform a series of 2D and 3D experiments using ^{15}N for filtering or measuring coupling constants, as well as a series of experiments for investigating backbone dynamics.

From the spectra, more than 500 long range NOEs, 75 dihedral angles and, using amide exchange rates, 33 hydrogen bonds were identified and used as structural constraints. The final structure has an RMSD of ca. 0.4 Å on the protein backbone and more than 70% of the residues are located within the most favoured regions of the Ramachandran plot. The global fold of the current structure has an apparent similarity to earlier reported plastocyanin structures.

The dynamics experiments reveal a very flexible loop in the northern part of the protein as well as increased dynamics in the eastern acidic patches. The flexibility in these regions is probably associated with the function of plastocyanin in protein-protein interaction.

NMR CONFORMATIONAL STUDY OF THE CYTOPLASMIC DOMAIN OF THE CANINE Sec61γ PROTEIN FROM THE PROTEIN TRANSLOCATION PORE OF THE ENDOPLASMIC RETICULUM MEMBRANE

Véronica Beswick, Françoise Baleux^, Tam Huynh-Dinh^, François Képès[+], Jean-Michel Neumann and Alain Sanson

Département de Biologie Cellulaire et Moléculaire, Section de Biophysique des Protéines et des Membranes, URA CNRS 2096, and [+]Service de Biochimie et de Génétique Moléculaire, CEA Saclay, 91191 Gif sur Yvette Cedex, France; ^Unité de Chimie Organique, URA CNRS 487, Institut Pasteur, 28 rue du Dr Roux, 75724 Paris Cedex 15, France

Conformational studies of the synthesized N-terminal cytoplasmic domain of the canine Sec61γ protein, an essential protein from the translocation pore of secretory proteins across the endoplasmic reticulum membrane, were performed using two dimensional proton NMR spectroscopy. This canine domain is one of the smallest domains within the homologous protein family and may thus constitute the minimal functional structure. The peptide was solubilized in pure aqueous solution or in the presence of dodecylphosphocholine micelles mimicking a membrane-solution interface. In pure aqueous solution, the peptide is remarkably unfolded. Forming a stable complex with dodecylphosphocholine micelles, it acquires a well defined α helix-loop-α helix secondary structure, with the first helix, highly amphipathic, lying at the micelle surface. The loop comprising four residues, is delimited by two flanking helix capping structures, highly conserved in the whole homologous protein family. No tertiary structure, that could have been revealed by inter-helix NOE contacts, was observed. From these experimental results and using general arguments based on sequence information and knowledge on peptide-membrane interactions, a structure of the entire Sec61γ protein in membrane bilayers is proposed.

FUNCTIONAL DIVERSITY OF PH DOMAINS: AN EXHAUSTIVE MODELING STUDY

Niklas Blomberg and Michael Nilges

Structural Biology, European Molecular Biology Laboratory, D-690 12 Heidelberg, Germany

Pleckstrin homology (PH) domains are found in a large number of intracellular protein involved in signalling or cytoskeletal organisation [2]. Several recent structures of PH-domains have shown that despite the low sequence similarity (20% average pairwise identity) this family has a well conserved structural core. The general function of this domain is unclear but it has been shown that some PH domains bind to phospholipids or phospholipid head groups [1]. Together with the strong electrostatic polarisation seen in the experimental structures, this has led to speculations about a membrane localising function for this domain. We have analysed the electrostatic properties and the spatial amino acid distribution for this family using automatically generated homology models.

An evaluation of the models of known structures shows that the core of the proteins are well modeled in spite of the low homology and that most of the deviations are found in the long loop regions. Even though the loops carry most of the charges in the PH domain family it is found that the low precision of the loops had very little influence on the global charge properties. We find that most PH domains show the same electrostatic properties as the known structures. However, there is a group of domains with very different electrostatic properties, in which most of these domains have a Cdc24-homology domain immediately on the N-terminal side of the PH domain. We also find that in proteins with two internal PH domains there is a electrostatic complementarity between these.

The electrostatic analysis of the PH-domain models shows that charge polarisation is a general property of this family. The majority of the PH domains show a polarisation similar to the solved experimental structures, strengthening the idea that most PH-domains bind phospholipid membranes. Our finding of a subgroup of PH-domains with an opposite polarity raises the question if phospholipid binding is a general function for the PH domain or whether there are other ligands yet to be found. We also demonstrate a use of homology modelling as an elegant general sequence analysis tool which yields significantly more information than conventional alignment analysis.

REFERENCES

1. K. M. Ferguson, M. A. Lemmon, P. B. Sigler, and J. Schlessinger (1995) *Nature Struct. Biol.* 9, 715–718.
2. M. Saraste and M. Hyvoenen (1995) *Curr. Op. Struct. Biol.* 5, 403–408.

STRUCTURE OF AN ESSENTIAL REGION OF THE P53 TRANSACTION DOMAIN

Maria Victoria E. Botuyan,[1] Jamil Momand,[2] and Yuan Chen[1]

[1]Division of Immunology & [2]Department of Cell and Tumor Biology, City of Hope National Medical Center, Duarte, CA 91010

The peptide segment surrounding residues Leu 22 and trp 23 of the p53 protein plays a critical role in the transaction activity of p53. This region binds basal transcriptional components such as the TATA box-binding protein Associated Factors $TAF_{II}40$ and $TAF_{II}60$ as well as the mdm-2 and Adenovirus type 5 E1B 55 kD oncoproteins. The structure of residues 14 though 28 of p53 has been studied by nuclear magnetic resonance spectroscopy, and has been found to contain two β–turn structures with the functionally critical residues Phe 19, Leu 22, Trp 23 and Leu 25 forming a hydrophobic cluster. The positional pattern of these hydrophobic residues is conserved among several transcriptional activators. This structural propensity observed in p53 may represent a common motif within transactivation domains required for binding components of the basal transcriptional machinery.

AN INVESTIGATION INTO THE STRUCTURE AND INTERACTIONS OF UPSTREAM BINDING FACTOR

Janice Bramham, Brian McStay and David G. Norman

Department of Biochemistry, Medical Sciences Institute, University of Dundee, Dundee DD1 4HN, Scotland, United Kingdom

The activation of transcription of eukaryotic ribosomal RNA genes by RNA polymerase I (pol I) requires the presence of two trans-acting factors for initiation, namely upstream binding factor (UBF) and SL1. These dedicated transcription factors bind to the ribosomal gene promoter thereby forming a stable pre-initiation complex which is specifically recognised by pol I and which supports multiple rounds of transcription initiation.

UBF is the archetypal member of the large family of transcription factors which utilise multiple HMG (high mobility group) domains to bind to DNA. It has become apparent that UBF acts through a series of protein-DNA and protein-protein interactions. In order to investigate the structure and interactions of UBF, this work is focussing upon the structures of the N-terminal dimerisation domain and the HMG box domains, in particular HMG box 1 of *Xenopus* UBF, using heteronuclear, high-resolution NMR spectroscopy and moecular modelling. Unlabelled and labelled (^{15}N) protein samples have been purified from recombinant, poly-histidine fusion proteins after over-expression in *E. coli*. Modelling work on the HMG domains has taken advantage of the NMR-derived solution structures of HMG domains from a number of proteins.

CONFORMATIONAL STUDIES OF GLUCAGON-LIKE PEPTIDE-1(7–36)-AMIDE IN 2,2,2-TRIFLUOROETHANOL/WATER BY NMR

Xiaoqing Chang and Jens J. Led

Department of Chemistry, University of Copenhagen, The H. C. Ørsted Institute, Universitetsparken 5, DK-2100 Copenhagen Ø, Denmark

The active products of the proglucagon peptide form one branch of the growth hormone-releasing factor superfamily of peptides. Among them, glucagon-like peptide-1-(7–36)-amide (GLP-1) is a potential substitution in curing the non-insulin-dependent diabetes mellitus. The conformational properties of glucagon depend strongly on the solution conditions. In pure aqueous solution, GLP-1 is known as a random coil; no secondary structure has been detected. However, by adding small organic molecules like 2,2,2-trifluoroethanol (TFE), GLP-1 can gain an entirely helical structure. This dramatic change is of great significance, since the active form of GLP-1 is known as a helix. Detailed conformational studies of GLP-1 in TFE/water mixture have been carried out by NMR. It is shown that GLP-1 gradually gains a stable helical secondary structure upon addition of TFE up to around 40% (v/v). Furthermore, the provisional three-dimensional solution structure of GLP-1 in 40% TFE has been determined by 2D NMR and restrained molecular dynamics calculations. The comparison of the conformational properties of GLP-1 in TFE and its native state in pure aqueous solution can provide further insights into the folding process of this peptide.

ANNEXIN FOLDING: NMR CONFORMATIONAL STUDY OF ANNEXIN ISOLATED DOMAINS

Françoise Cordier-Ochsenbein[#], Raphaël Guerois[‡], Françoise Russo Marie[¥], Pierre Noël Lirsac,[†] Jean-Michel Neumann[‡] et Alain Sanson[‡+]

[#] Centre CNRS, ICSN, Laboratoire de RMN, 91190 Gif sur Yvette; [‡] DBCM, SBPM, URA CNRS 2096, CEA Saclay, 91191 Gif sur Yvette, Cedex; [†] DIEP, CEA Saclay, 91191 Gif sur Yvette, Cedex; [¥] ICGM, Unité INSERM 332, 22, rue Méchain, 75014 Paris.[+] Also from PMC University, 9 Quai Saint-Bernard, Bât. C, 75005 Paris, France

Annexins constitute an interesting protein model for folding studies, because of their highly hierarchic structure: four consecutive homologous domains assembled into two modules. In the native structure, each domain (~ 70 residues) exhibits the same topology, *i.e.* five helices folded into a characteristic super-helix motif: (A-loop-B)-C-(D-E). Using NMR and CD spectroscopies, the study of the annexin folding was undertaken, by analyzing the residual structures in isolated domains or modules. Remarkably, annexin I domain 1 was shown to constitute independent folding units, whereas annexin I domain 2 exhibits a secondary structure amount of about 30% of the native secondary structure. Moreover, annexin V domain 1 was also able to fold independently, indicating that this property could be a general feature of the annexin family. Taking all our results in account, including experimental data concerning annexin I domain 1, annexin V domain 1, annexin I domain 2, and the 2–3 module of annexin I, we propose a sequential model for the folding of annexin I.

The partly folded annexin I domain 2 can be considered as an equilibrium unfolded state which contributes to the folding process of the annexin I protein. A complete analysis of the residual structure of annexin I domain 2 was thus undertaken using ^{15}N-^{1}h 3D-NMR. The residual structures can be separated into two sets: a set of native secondary structures and a set of non-native local structures. The set of native secondary structures consists of three regions with large helical populations and in rather sharp correspondence with the native A, B and E helices. A small population of native helix structure is also observed in the second part of the C-helix. The set of non-native local structures comprises several helix-disruptive side-chains, generally aspartate side-chains, to backbone interactions, a localized network of side-chain to side-chain interactions or salt-bridges. This ensemble of non-native local interactions stabilize the unfolded state of the domain at the expense of the folded state. Interestingly, important residues involved in these local interactions in the isolated domain are involved in crucial long range side-chain to side-chain interactions in the native state. This implies a switching from local and unfolded state-stabilizing interactions to non-local folded state-stabilizing interactions during the folding process. This supports the hypothesis that the residues involved in such a switching interaction may be part of the sequence-coded devices that structurally and kinetically control the ensemble of folding pathways of the proteins.

SOLUTION STRUCTURE BY NMR OF CIRCULIN A: A MACROCYCLIC PEPTIDE HAVING ANTI-HIV ACTIVITY

Norelle L. Daly, Anita Koltay, Kirk Gustafson[#] and David J. Craik

Centre for Drug Design and Development, University of Queensland, Brisbane, QLD 4072, Australia and Victorian College of Pharmacy, Monash University, 381 Royal Parade, Parkville, VIC 3052, Australia and [#]National Institutes of Health, National Cancer Institute, Frederick Cancer Research and Development Centre, Frederick, Maryland 21702–1201, USA

The three-dimensional solution structure of circulin A, a 30 residue polypeptide from the African plant *Chassalia parvifolia*, has been determined using 2D ^{1}H NMR spectroscopy. Circulin A was identified because it inhibits the cytopathic effects and replication of the human immunodeficiency virus[1]. Structural restraints consisting of 361 interproton distances inferred from NOEs, and 8 backbone dihedral angle restraints from spin-spin coupling constants were used as input for simulated annealing calculations and energy minimisation in the program X-PLOR. Circulin A adopts a compact structure consisting of β-turns and a distorted segment of triple-stranded β-sheet. The molecule is stabilised by three disulfide bonds, two of which form an embedded loop completed by the backbone fragments connecting the cysteine residues. A third disulfide bond threads through the centre of this loop to form a cysteine-knot motif. This motif is present in a range of other biologically active proteins including ω-conotoxin GVIA and Cucurbita maxima trypsin inhibitor (CMTI-I)[2]. Circulin A also belongs to a novel class of macrocyclic peptides which have been isolated from plants in the Rubiaceae family. The global fold of circulin A is similar to kalata B1, the only member of this class for which a structure has previously been determined.

1. Gustafson et al. (1994) *J. Am. Chem. Soc. 116*, 9337–9338.
2. Pallaghy et al. (1994) *Protein Science 3*, 1833–1839.

CHEMICAL SHIFT CALCULATIONS: A NEW TOOL FOR MACROMOLECULAR STRUCTURE DETERMINATION

A. Dejaegere, F. Sirockin, M. Karplus, D. Sitkoff, D. Case, and J-F. Lefèvre

Ecole Superieure de Biotechnologie de Strasbourg, Groupe RMN, Bd. S. Brant, 67400 Illkirch, France

NMR is an important technique for structure determination of proteins and nucleic acids. The large number of NMR structures now available has made possible the development of semi-empirical theories for chemical shift dispersion that allow the calculations of chemical shifts in proteins. We used a semi-empirical model to calculate the chemical shifts for PMP-D2 and ω-conotoxin MVIIA; both are small proteins (35 and 25 residues, respectively) with three disulfide bridges. Some regions of the proteins showed large discrepancies between calculated and experimental shifts. Heteronuclear relaxation studies subsequently showed that these discrepancies are due to slow conformational exchange in the molecule. These results showed that chemical shift calculations can serve as a useful independent test of the quality of the NMR structures.

Nucleic acid structure determination by NMR often suffers from a scarcity of experimental NOE restraints; chemical shift information should help in structure refinement. We are developing a model for nucleic acids that combines semi-empirical and density functional calculations of chemical shifts. The model shows that chemical shifts are sensitive to details of nucleic acids structure.

COMPARATIVE ASSESSMENT OF THE QUALITY OF NMR STRUCTURES: A STATISTICAL SURVEY

J. F. Doreleijers, J. A. C. Rullmann and R. Kaptein

NMR Spectroscopy, Utrecht University, The Netherlands

Over the last five years the number of macromolecular structures determined with NMR spectroscopy has increased dramatically. The Protein Data Bank (PDB) [1] currently contains more than 700 coordinate entries of NMR origin, i.e., 14 percent of the total number of entries. These structures have been obtained using a wide variety of methods and parameter sets. Some comparative studies of protocols have appeared in the literature.

These focused on the precision of structure determinations and the effect of adding more restraints, but did not pay much attention to other quality indicators, such as molecular geometry and the agreement between structures and experimental data. A recent conformational study comprised 22 ensembles and 9 single structures [2]. Here we present results of a comprehensive analysis of 101 protein coordinate sets. This study is part of a larger programme aimed at validation of biomolecular structures determined with X-RAY or NMR spectroscopy [3].

1. F. C. Bernstein, T. F. Koetzle, G. J. Williams, E. F. Meyer Jr, M. D. Brice, J. R. Rodgers, O. Kennard, T. Shimanouchi and M. Tasumi (1977) *J. Mol. Biol. 112*, 535–542.
2. M. W. MacArthur, R. A. Laskowski and J. M. Thornton (1994) *Cur. Opinion Struct. Biol. 4*, 731–737.
3. On World Wide Web: http://biotech.embl-ebi.ac.uk:8400/.

STRUCTURAL AND DYNAMIC CHARACTERIZATION OF DENATURED LYSOZYME

Klaus M. Fiebig,[1] Harald Schwalbe,[2] Lorna J. Smith,[3] and Christopher M. Dobson[4]

[1]Department of Molecular Biophysics & Biochemistry, Yale University, New Haven, CT 06520; [2]Instutut für Organische Chemie, Universität Frankfurt/Main, Marie-Curie-Strasse 11, 60449 Frankfurt/Main, Germany; [3]Oxford Centre for Molecular Sciences, New Chemistry Laboratory and [4]Inorganic Chemistry Laboratory, University of Oxford, Oxford OX1 3QR, United Kingdom

Urea denatured hen lysozyme in its oxidized and reduced forms has been characterized by heteronuclear NMR techniques. Near complete sequential assignment could be obtained using ^{15}N filtered NOESY and TOCSY experiments. Continuous stretches of (i,i+1) and a multitude of sequential (i,i+2) and (i,i+3) NOEs were extracted from the NOESY spectra. Additionally, $^3J(HN,H_\alpha)$ coupling constants and amide relaxation rates were measured. This multitude of experimental data was analyzed in terms of theoretical random coil models. Coupling constants and NOE intensities agree surprisingly well with theoretical predictions generated by a random coil polypeptide model based on intrinsic ϕ,ψ properties found in the protein data bank. This result suggests that local conformational preferences of denatured proteins are predominantly determined by nearest neighbor interactions along the peptide chain. Support for this hypothesis also stems from the observation that the four disulfide cross-links in the oxidized protein do not significantly perturb chemical shifts, coupling constants, and NOEs when compared to the reduced protein. However, the dynamic properties of denatured lysozyme differ significantly in its reduced and oxidized form. Analysis of the relaxation data in terms of a simple random coil dynamics model provides evidence for local hydrophobic clustering near tryptophan residues and increased conformational barriers due to disulfide crosslinking. This study argues that urea denatured lysozyme is an ensemble of conformers with little, if any, local strucure but complex global interactions originating from disulfide bond constraints and weak hydrophobic forces.

ROLE OF PROTEIN DYNAMICS IN ELECTRON TUNNELING IN PHOTOSYNTHETIC REACTION CENTER

E. N. Frolov, V. Goldanskii and F. Parak

Institute of Chemical Physics, Russian Academy of Sciences, Moscow region, 142432, Chernogolovka, Russia

Mössbauer spectroscopy has been used to investigate the temperature dependence of mean square displacements of iron atoms in the photosynthetic protein "reaction center" from *Rps. viridis*. It was shown that the freezing of the protein dynamics displayed by the Mössbauer effect correlates with the efficiency of the electron tunneling (ET) from the proximal heme iron of the cytochrome subunit to the cation-radical of bacteriochlorophyll (P). The nature of this correlation is considered to be a result of the formation of the out-of-equilibrium state of P under its photoinduction at low temperature. The electronic level of this state differs from its equilibrium state in the value of reorganization energy of P formation and is equal to 0.13 eV, which is higher than the free energy of reaction, equal to 0.12 eV at room temperature. It leads to the energetic ban which blocks tunneling. Under reduction of the neighboring heme(s), the energetic ban vanishes due to the electrostatic shift of the electronic level of the heme-donor and ET is reinstated. The value of the electrostatic shift of the electronic level of the heme-donor is calculated as equal to 0.04 eV and 0.11 eV under reduction of the heme groups located at 7 and 11 Å respectively. An empirical relation for the rate of ET is given.

SOLUTION STRUCTURE OF WILD GRB2 N-SH3 DOMAIN AND ITS MUTANT TYR7 → VAL7

Edith Gincel[1], Nathalie Goudreau[1], Michel Vidal[1], Fabrice Cornille[1], Christiane Garbay[1], Febienne Parker[2], Marc Duchesne[2], Brune Tocque[2] and Bernard-P. Roques[1]

[1]Département de Pharmacochimie Moléculaire et Structurale U266, INSERM-URA D1500 CNRS, Universite Rene Descartes-Paris V, 4 Avenue de l'Oservatoire, 75270 Paris Cedex 06 and [2]Rhone Poulenc Rorer, Centre de Vitry-Alfortville, 13 quai Jules Guesde, 94403 Vitry/Seince Cedex, France

Grb2 is a small adaptor protein of 217 amino acids, comprising one SH2 domain surrounded by two SH3 domains. Grb2 couples receptor tyrosine kinase activation to Ras signalling by interacting, through its SH3 domains, to the carboxy-terminal proline-rich region of the guanine nucleotide exchange factor Sos.

The structure of the amino-terminal SH3 domain of Grb2, (in which the Cys32 was mutated to a Serine to stabilize the complex) complexed with a proline-rich peptide derived from Sos: VPPPVPPRRR, was determined by NMR and molecular modeling. The peptide adopts a left-handed polyproline type II helix conformation. The folding of the N-SH3 consists of 2 three stranded antiparallel β−sheets packed against each other at approximately right angles, with one sharing strand that begins in one sheet and continues in the other, resulting in a β barrel-like structure. Two subsites are present: S1, consisting of the side chains of Y7 and Y52, interacting with P2 of the peptide and S2, consisting of the side chains of F9, W36, Y52, forming an hydrophobic core in which V5 of the peptide can be inserted.

Several mutants of this N-SH3 domain were studied to understand the way of interactions: 1) P49 → L49, 2) E40 → T40, 3) Y7 → V7. The first mutation induces a loss of peptide affinity. The complex (studied by NMR) is not structured. The second was done in the interaction site with the P97 protein. The complex was very similar to the S32-N-SH3 domain with VPPPVPPRRR. The last mutation was supposed to induce a total loss in affinity to the peptide (as was seen for other homologous complexes) but it was only 10 times reduced. It was studied by molecular modeling based on NMR data.

Actually, the global forms of the protein are the same, but the peptide rotates and presents another face to the protein.

HETERONUCLEAR RELAXATION STUDY OF β-SPECTRIN PH DOMAIN AND ITS COMPLEX WITH INOSITOL-1,4,5-TRIPHOSPHATE: EVIDENCE OF RESTRICTED LOOP MOTION UPON LIGAND BINDING

Michael R. Gryk, Roger Abseher, Bernd Simon and Hartmut Oschkinat

European Molecular Biology Laboratory, Meyerhofstrasse 1, D-69117 Heidelberg, Germany

In several ways the pleckstrin homology (PH) domain of mouse brain β-spectrin is a biological NMR spectroscopist's dream come true. The protein provides such highly dispersed and highly resolved spectra that the NMR structure was able to be determined resorting solely to homonuclear techniques [1]. Despite the resolution of the spectra and the benefits it has afforded, two residues in the loops responsible for recognition of the PH domain's natural ligand, a phosphatidylinositol lipid, could not be assigned. The resulting appearance of these residues in the spectra of an aqueous solution of PH domain containing inositol-1,4,5-triphosphate (IP3), an analog for the water-soluble portion of its lipid target, together with the sharpening of other resonances in the ligand binding loops [2] led to the speculation that the loops responsible for ligand binding were highly mobile in the free state but adopt a more rigid conformation in the bound state. As such, broad lines in the free state could be rationalized as being due to chemical shift averaging, the sharpening of the lines in the bound state signifying a concomitant restriction of motion.

We report here heteronuclear relaxation measurements which support this claim. We have measured the ^{15}N T_1 and T_2 relaxation rates as well as the ^{1}H-^{15}N heteronuclear NOE for the backbone amide groups of both uniformly ^{15}N enriched PH-domain, and [U-^{15}N] PH-domain bound to IP3. Three distinct regions of the protein appear to be mobile in this ^{15}N relaxation study - the β1-β2 and β5-β6 turns responsible for binding IP3 and the carboxy terminal residues 104–106. Upon addition of the IP3 ligand, motion of the binding loops is decreased as inferred from the observed differences in the heteronuclear relaxation rates. Comparison with the recent ^{15}N relaxation study of the dynamic PH domain [3] will be made.

1. M. J. Macias, A. Musacchio, H. Ponstingl, M. Nilges, M. Saraste and H. Oschkinat (1994) *Nature 369*, 675–677.
2. M. Hyvonen, M. J. Macias, M. Nilges, H. Oschkinat, M. Saraste and M. Wilmanns (1995) *EMBO J. 14*, 4676–4685.
3. D. Fushman, S. Cahill and D. Cowburn (1997) *J. Mol. Biol. 266*, 173–194.

NATIVE AND NON-NATIVE CAPPING BOXES AS POSSIBLE FOLDING PATHWAY DETERMINANTS IN HELICAL PROTEINS: AN EXAMPLE FROM ANNEXIN FRAGMENTS

Raphaël Guerois, Benoît Odaert, Françoise Baleux,[⊥] Tam Huynh-Dinh,[⊥] Jean-Michel Neumann and Alain Sanson.

Département de Biologie Cellulaire et et Moléculaire, SBPM, URA CNRS 1290, CEN Saclay, 91191 Gif sur Yvette, Cedex, France, and [⊥]Unité de Chimie Organique, URA CNRS 487, Institut Pasteur, 28 rue du Dr Roux, 75724 Paris, Cedex 15, France

Understanding how a polypeptide chain folds into a particuliar 3D structure remains a central interest in biochemistry. Little is known about the structure and dynamics of the intermediate states that prevail during the first few milliseconds of the folding process.

Protein fragments, which have lost all or part of their long range interactions leading normally to a cooperative folding, can be used as pertinent indicators of the local early state of the protein folding. Fragments spanning elements of secondary structure are important for understanding initiation, propagation and stability of these structures. In addition, step by step increase of the fragment length may be a valuable means for understanding how long range interactions take place to initiate the protein collapse into a compact state. We are currently using this approach for analysing folding and stability of a protein of the annexin family. These proteins are attractive because of their highly hierarchic structure: four ≈ 70 residue strongly homologous domains assembled into two modules. Each domain is made of five helices, successively named A to E, and folded into a characteristic super-helix motif: (A-loop-B)-C-(D-E).

Five fragments, ranging from ≈ 20 to ≈ 70 residues in length, thus spanning single helices up to the full domain 2 of human annexin I, have been synthesized. These fragments were studied using 2D high resolution NMR.

Here we will focus on the (A) fragment spanning the A helix of Annexin I domain 2:

$$\text{Ac-} \quad \underset{1}{L} \ K \ \underset{}{T} \ P \ A \ \underset{5}{Q} \ F \ \underset{}{D} \ A \ \underset{10}{D} \ \underset{}{E} \ L \ R \ A \ A \ \underset{15}{M} \ K \ G \ \text{-(NH}_2)$$

$$\llcorner < \text{————————} A \ helix \text{————————} > \lrcorner$$

which was obtained by chemical synthesis. This fragment indeed includes several interesting features whose study could help characterizing some aspects of the folding pathway of the protein:

a. Two potential capping boxes "T_3xxQ_6" and "D_8xxE_{11}". The former corresponds to the *native* hydrogen bonds network stabilizing the A helix amino end. The latter is a *non-native* capping box which breaks the helix at residue D_8.

b. A "$D_{10}xxR_{13}$" salt bridge and other features associated with the hydrophobic residues.

In aqueous solution, several populations of conformers can be distinguished. The first one presents a helical conformation from residue A5 to M16. The other one is also helical, but the helix is broken at residue D8. Protein fragments, studied in aqueous solution, present very dynamic structures exchanging at high rate compared to the NMR time scales. Our aim was to describe the various families of conformers adopted by the peptide, and the importance of each local structures in the stability of helical conformations. An attempt of quantitative analysis will be presented. In addition, the study of eight punctual mutants have been carried out in order to analyse the role of each residue in the ocurrence of these residual structures.

This analysis will stress the relationship existing between the sequence and the folding pathway of the protein. The non-native "D_8xxE_{11}" capping box, which is of course dismantled in the final protein structure, involves highly conserved residues among the annexin family. Since these residues are implicated in long range interactions in the native structure, one may suggest the ocurrence of a switching from local non native interaction to the native conformation during the folding process. We postulate that the double stop signal present in A helix may act as a compacting "helper" and plays an important role in the folding pathways and kinetics.

ON THE RESOLUTION OF AMBIGUOUS NOEs

Thomas J. Hoeffel and Timothy S. Harvey

Amgen Inc., 1840 DeHavilland Drive, Thousand Oaks, CA 91320

The development of multi-dimensional heteronuclear NMR using isotopic enrichment has made the structure determination of increasingly large proteins (> 25 kDa) tractable. However, increases in dimensionality as a result of isotopic editing are limited by the finite chemical shift dispersion as the size of the system increases. Since the use of interatomic distance restraints requires the unique identification of the ^1H chemical shifts giving rise to an NOE, these may not be uniquely determined *a priori* in large molecules. Thus, the spectral overlap still posses one of the most vexing, error prone, and labor intensive stages in the structure determination process. To address this, methods have been developed which rely less on the subjective interpretation of data, and more on rigorous statistical techniques. The strength of this approach is the nearly complete automation of the refinement, greatly reducing the inherent error and time associated with such a task.

A protocol incorporating ambiguous distance restraints has been developed which allows the automated analysis of 3- and 4-dimensional NOE data sets. Using an simple, iterative, and statistically based method, ambiguous NOEs are systematically reduced to simple pairwise restraints when appropriate. In addition, this procedure may be used to identify erroneous assignments as well as spectral artifacts. Examples are given which illustrate the benefits and pitfalls of this paradigm.

NMR STUDIES OF *trp* REPRESSOR: DNA INTERACTIONS

Mark Jeeves

School of Biochemistry, University of Birmingham, Edgbaston, Birmingham, B15 2TT, United Kingdom

In order to understand the specificity of the *E. coli trp* repressor for its operators, we have begun to study complexes of the protein with alternative DNA sequences, using ^{1}H NMR spectroscopy. We have looked at two twenty base pair oligodeoxynucleotides: one a symmetrised form of the *trpR* operator (*trpR^S*) and the other a symmetrised form of a mutant of the *trpO* operator (*trpO^M*). Deuterated protein was used to assign the spectrum of the *trpR^S* oligodeoxynucleotide in a 37 kDa complex with the *trp* holorepressor. Many of the resonances of the DNA shift on binding to the protein, suggesting changes in conformation throughout the sequence. The largest changes in shifts for the aromatic protons in the major groove are for A15 and G16, which are thought to hydrogen bond to the protein, possibly via water molecules. We have also attempted to calculate the structure of the DNA, both free and bound, in order to study the effect of DNA sequence on structure and also to see if there are any differences in structure which may account for the lower affinity of the *Trp* repressor for the mutant DNA. Also difference in the HSQC protein spectra for the two complexes suggests a difference in the orientation of the recognition helix.

STRUCTURAL ANALYSIS OF THE RECEPTOR ASSOCIATED PROTEIN

[1]Peter Holme Jensen, [1]Peter Reinholt Nielsen, [2]Lars Ellgaard, [2]Michael Etzerodt, [2]Hans C. Thøgersen and [1]Flemming M. Poulsen

[1]Carlsberg Laboratory, Dept. of Chemistry, Gamle Carlsberg Vej 10, 2500 Valby, Denmark; [2]Laboratory of Gene Expression, Dept. of Molecular and Structural Biology, University of Aarhus, Gustav Wieds Vej 10, 8000 Aarhus C., Denmark

The Receptor Associated Protein (RAP) is 39 kDa glycoprotein and consists of 323 amino acids. It is known to bind to different cell surface endocytosis receptors, such as the α_2-macroglobulin receptor (α_2-MR), the glycoprotein 330 (gp330), the low density lipoprotein receptor (LDLR) and the very low density lipoprotein receptor (VLDLR), all receptors being part of the LDLR family. Through this binding, RAP functions as a competitive inhibitor of all ligands known to bind to the LDLR family receptors (α_2-MR, gp330, LDLR and VLDLR). RAP is in mammals located in the endoplasmatic reticulum and in the GOLGI-apparatus. It consists of three domains. RAP domain 1 (RAP D1) comprises 96 amino acids. A binding site in RAP is located to the C-terminus part of D1. The solution structure of a truncated part of RAP D1 (81 aa), not including the binding site, has been elucidated by Peter Reinholt Nielsen. The full RAP D1 has also been target for a structural analysis using heteronuclear multidimensional NMR. This showed no indication of secondary or tertiary structure in the binding site region. Structural analysis of RAP Domain 1 & 2 together (201 aa) is now being performed. It is the hope that this analysis will give the solution structure of D2, and indicate whether or not the C-terminal part of D1 becomes structured when RAP D1 is expressed together with D2.

BIOPHYSICAL STUDIES OF THE ALBUMIN-BINDING GA MODULE

Maria U. Johansson[1], Lars Björck[2], Torbjörn Drakenberg,[1] and Sture Forsén[1]

[1]Department of Physical Chemistry 2, Chemical Center, Lund University, P.O. Box 124, S-221 00 Lund, Sweden and [2]Department of Cell and Molecular Biology, Section for Molecular Pathogenesis, Lund University, P.O. Box 94, S-221 00 Lund, Sweden

The results presented in this poster is a biophysical approach to gain knowledge about the structure, stability and mode of interaction of the albumin-binding GA module of the bacterial surface protein PAB from *Peptostreptococcus magnus*. Using CD spectroscopy we have shown that the GA module is very stable with respect to temperature and pH variations. The GA module has a rather unchanged secondary structure between pH 2 and 11 and contains 60% and 27% α-helix at 27 and 95°C, respectively. The module starts to unfold above 70°C, but due to gradual unfolding over a large temperature range it was not possible to determine the the transition midpoint of the thermal denaturation. The melting process seems reversible. During the sequential assignment procedure an error in the published amino acid sequence was discovered. The secondary structure elements were identified by sequential and medium range NOEs, values of coupling constants, chemical shift indices and the presence of slowly exchanging amide protons. The global fold was shown to be similar to the IgG-binding domains of staphylococcal protein A based on a few key NOEs. The rms deviation between the two three-helix bundles after the final round of structure calculation was around 3 Å despite the fact that the sequences between these two functionally different domains are unrelated. The rotational correlation time was 4.35 ns when the relaxation behaviour of the ring NH proton of Trp5, following either a non-selective or a selective 180° pulse, was analysed by comparing it to the full relaxation curves. A theoretical 2D NOESY spectrum was back-calculated from the 20 structures in the final ensemble. In order to identify amino acid residues in the GA module interacting with albumin we have carried out hydrogen-deuterium exchange experiments of the albumin-bound GA module following the technique developed by Paterson et al. (1990). However, it was not possible to identify a distinct albumin-binding surface by comparing the hydrogen-deuterium exchange rates for the free and bound GA module since nearly all residues in the complex are protected.

Paterson et al. (1990) *Science 249*, 755–759.

STRUCTURAL STUDIES OF A CHIMERIC ANTIAGGREGANT RGDX PEPTIDE BY PROTON NMR

Esther Kellenberger, George Mer, Gilles Travé and Jean-François Lefèvre

CNRS - UPR 9003, Ecole Superieure de Biotechnologie de Strasbourg, Bd Sebastien Brant, 67400 Illkirch-Graffenstaden, France

For a better understanding of the relationship between structure and function in cell-adhesion,we have designed a chimeric peptide containing the RGDX motif found in numerous proteins involved in these phenomena. We have modified the sequence of a scorpion toxin (leiurotoxin I) in order to maintain a RGDX sequence (QMIRGYFDV: sequence of the hypervariable loop of a monoclonal antibody able to inhibit platelet aggregation) in an helical conformation (a proton NMR study of the nonapeptide structure has suggested an helical conformational equilibrium).

The structure of leiurotoxin I, which consists of a C-terminus β-sheet that is flanked by a N-terminus α-helix, is stabilized by three disulfide bridges. The choice of leiurotoxin I has been determined by the fact that this α/β motif is also present in other toxins, even if there is no sequence homology (only the three half-cystine and a glycine are conserved).

The structural feature of the chimeric peptide, called LI-RGD, has been investigated by standard two-dimensional NMR techniques. Complete sequence-specific assignments of the individual backbone and side-chains resonances were achieved using through-bond and through-space connectivities. The structure calculations were performed from NOE-derived interproton distances, dihedral angles calculated from JHN-H$_\alpha$ coupling constants and hydrogen bonds deduced from proton-deuteron exchange rates.

LI-RGD structure closely resembles that of leiurotoxin I. LI-RGD and the QMIR-GYFDV peptide inhibit platelet aggregation with the same efficiency.

STRUCTURAL AND FUNCTIONAL STUDIES OF INSULIN GENE ENHANCER BINDING PROTEIN Isl-1

G. Larsson, P. O. Lycksell, and G. Behravan

Dept. of Medical Biochemistry and Biophysics, Umea University, Sweden

Purpose. Isl-1 is involved in the regulation of gene expression of the insulin, amylin and proglucagon genes. Isl-1 has a molecular weight of 42 kDa. A domain of about 70 amino acid residues, related to homeobox regions in other DNA binding proteins, can be distinguished. Isl-1 also contains two metal binding Cys/His motifs, denoted as LIM domains. This study addresses the question of how the protein recognizes its target DNA, and what kind of driving forces are present in specific DNA binding.

Methods. The homeodomain and LIM domain with two Cys/His motifs, as well as the intact protein were cloned and overexpressed. DNA binding of the homeodomain was monitored by gel shift assay, BIAcore and NMR. Structural properties for the same domain were studied by CD-spectroscopy and NMR. Thermal and chemical unfolding of homeodomain in the presence and absence of DNA was studied by following the change in the CD-signal for α-helices.

Results and Conclusion. Soluble protein could be expressed, and purification protocols for all domains, as well as for the intact protein, have been developed. Specific DNA binding of the homeodomain was demonstrated. The homeodomain has a high content of α-helices. When the homeodomain binds to its target DNA, an increase in stability is observed. Preliminary NMR data shows that the homeodomain has a well defined structure. NMR data also shows that the homeodomain binds to a -TAAT- core sequence in the DNA.

BARLEY LIPID-TRANSFER PROTEIN COMPLEXED WITH PALMITOYL CoA: THE STRUCTURE REVEALS A HYDROPHOBIC BINDING SITE WHICH CAN EXPAND TO FIT BOTH LARGE AND SMALL LIPID-LIKE LIGANDS

Mathilde H. Lerche, Birthe B. Kragelund & Flemming M. Poulsen

Carlsberg Laboratory, Department of Chemistry, Gamle Carlsberg Vej 10, DK-2500 Valby, Denmark

Plant non-specific lipid-transfer proteins (nsLTPs) bind a variety of very different lipids *in vitro*, including phospholipids, glycolipids, fatty acids and acyl Coenzyme As. It is of interest to understand the structural mechanism of this broad specificity and its relation to the *in vivo* function of nsLTPs. In this study we have determined the structure of the complex between a nsLTP isolated from barley seeds (bLTP) and the ligand palmitoyl coenzyme A (PCoA) by ^{1}H and ^{13}C nuclear magnetic resonance spectroscopy (NMR). The palmitoyl chain of the ligand was uniformly ^{13}C-labelled allowing for the two ends of the hydrocarbon chain to be assigned. The four-helix topology of the uncomplexed bLTP is maintained in the complexed bLTP. The bLTP binds only the hydrophobic parts of PCoA with the rest of the ligand remaining exposed to the solvent. The palmitoyl part of the ligand is placed in the interior of the protein and bent in a U-shape. This part of the ligand is completely buried within a hydrophobic pocket of the protein. A comparison of the structures of the bLTP in the free form and in the ligated form suggests, that bLTP can accommodate long olefinic ligands by expansion of the hydrophobic binding site. This expansion is achieved by a bend of one helix, H_A and by conformational changes in both the C-terminus and helix H_C. This mode of binding is different from that seen in the structure of maize nsLTP (mLTP) in complex with palmitic acid, where binding of the ligand is not associated with structural changes.

STUDY OF THE PROTEIN DYNAMICS BY OFF-RESONANCE ROESY EXPERIMENT

T.E. Malliavin[¶], H. Desvaux[§], A. Padilla[¶] and M.A. Delsuc[¶]

¶ Centre de Biochimie Structurale, Faculté de Pharmacie, 15, av. Ch. Flahault, F-34070 Montpellier, France and § Laboratoire Commun de RMN, Service de Chimie Moléculaire, CEA, Centre d'Etudes de Saclay, F-91191 Gif sur Yvette Cedex, France

Quantitative analysis of molecular internal dynamics is usually performed by heteronuclear NMR. Nevertheless, the off-resonance ROESY (1) allows the analysis of protein dynamics by homonuclear NMR, through the observation of a relaxation phenomenon, which is the weighted sum of the NOESY and ROESY relaxations. The weights are determined by the angle θ between the B_0 and the B_1 spin-lock field. In this presentation, we are using this experiment to study internal dynamics of a protein. We propose to measure for each observed correlation, the ratio of the corresponding ROESY and NOESY relaxation rates, by recording several off-resonance ROESY spectra at different angles θ. These ratios can be related to the mobility of the vector connecting the hydrogens involved in the correlation.

The mobilities of more than 150 vectors connecting hydrogens inside a calcium-binding protein of 109 AA, the parvalbumin, were measured. The significance of the measured parameters was assessed by a careful analysis of error sources, and by a comparison with results obtained from ^{15}N and ^{13}C relaxation measurements (2). The protein backbone is found mainly rigid. The larger mobility found for some residues is consistent with independent experimental observations. The mobility found for some long-range vectors is analyzed in term of tertiary structure mobility.

1. H. Desvaux, P. Berthault, N. Birlirakis and M. Goldman (1994) *J. Magn. Res. Ser. A 108*, 219–229.
2. T. Alattia, A. Padilla and A. Cavé (1996) *Eur. J. Biochem. 237*, 561–574.

CONFORMATIONAL STUDIES OF PTH ANTAGONISTS DERIVED FROM PARATHYROID HORMONE RELATED PROTEIN CONTAINING LACTAM-BRIDGED SIDE-CHAINS

Stefano Maretto

Universita di Padova, Dipartimento di Chimica Organica, Centro di Studio sui Biopolimeri, Padova, Italy

Human parathyroid hormone (hPTH), an 84-residue linear peptide, is a major regulator of extracellular calcium homeostasis. It acts on receptors primarily located on bone and kidneys. The PTH-related protein (PTHrP) is a 141 residue secretory product released in consequence of pathological state such as humoral hypercalcemia of malignancy. Structure-activity relationship studies have demonstrated that the full active domain lies in the N-terminus portion (1–34) for both peptides. Further investigations yielded a novel class of antagonist peptides, PTH and PTHrP derivatives, based on the sequence 7–34. PTH(7–34) and PTHrP(7–34) share a very little sequence homology, nevertheless they are able to bind the same receptor, which suggests a similar bioactive conformation for the two fragments. These antagonists play a key role on the design of new drugs and they contribute to the elucidation of receptor structure and the mechanism of interaction.

Previous NMR studies led to the structural characterization of several PTH and PTHrP analogues, pointing to α-helix as principal secondary structure motif. On the hypothesis that helical conformation is largely responsible for the binding capability, three structurally constrained PTHrP analogues, containing (i)–(i+4) side-chain cyclization via lactam bridge, were designed.

1. [Lys13-Asp17, Lys26-Asp30]PTHrP(7–34)NH$_2$
2. [Lys26-Asp30]PTHrP(7–34)NH$_2$
3. [Lys13-Asp17]PTHrP(7–34)NH$_2$

The analogues still denote high antagonist activity and display strong propensity toward folding into helical conformation. CD investigations in aqueous solution and TFE-water mixtures showed that helix content increases with the percentage of TFE. NMR studies were carried out in 1:1 water-TFE mixtures and they generated interproton distances, used as constraints in the following structure calculations. We used the metric matrix distance geometry approach to search all conformational space and locate all structures which fulfill the NMR observables. The lowest energy structures from the DG calculations were submitted to Molecular Dynamics simulations: the peptides were soaked in a box of TFE in order to mimic the real sample conditions. During the simulations there is an extension and an enhancement of the helices with respect to the DG results, partially due to the addition of the partial charges within the MD force field. Furthermore, the utilization of the explicit solvent is necessary for accurate simulations.

The results from the conformational examination of these three analogs showed that the effect of the lactam bridge depends on location within the sequence. The 26–30 bridge produces similar results in both the mono and di-lactam containing peptides. This is to be expected given that this region is α-helical in the natural product. More importantly is the conformational consequences of the 13–17 lactam, that promotes the formation of helix locally but does not cause the continuation of the helix into the Arg19 region or the C-terminal region.

PRODUCTION OF PERDEUTERATED BIOMASS BY THE METHYLOTROPHIC YEAST *Pichia angusta* AND ANALYSES OF PHOSPHOLIPIDS

S. Massou, M. Tropis, F. Talmont, P. Demange, N. Lindley and A. Milon

Institut de Pharmacologie et de Biologie Structurale, IBCG-CNRS, 118 route de Narbonne, 31 062 Toulouse Cédex, France

The investigation of protein structure using modern spectroscopic techniques such as NMR, neutron diffraction require labelled compounds (^{13}C, ^{15}N, ^{2}H). Two strategies can be used to obtain labelled products; chemical synthesis and fermentation of micro-organisms. We have chosen the second approach because it allows the production of complex molecules such as proteins, lipids, steroids that cannot be synthesised.

We have grown the yeast strain *Pichia angusta* on deuterated media because it allows us to overexpress heterologous proteins using only methanol and water (1, 2). In order to optimise the growth of this micro-organism, we are studying the metabolic pathways which are inhibited during growth on deuterated media. ^{31}P and ^{1}H NMR is an efficient way of analysing phospholipid compositions. The total phospholipids were dissolved in a new solvent mixture composed of triethylamine, dimethyl-formamide, guanidium chloride and methanol. This solvent reduces linewidths and therefore allows investigation of the phospholipid compositions by using ^{31}P-^{1}H gradient multidimensional experiments. ^{31}P-NMR, with and without proton decoupling, and HMQC-HOHAHA gradient enhanced experiments have permitted unambiguous qualitative and quantitative phospholipid determination. We have thus shown that growth on deuterated media involves significant modification of phospholipid composition, which indicates the inhibition of specific metabolic pathways.

1. S. Haon, S. Augé, M. Tropis; A. Milon, N. Lindley. (1993) *J. Labelled Compnds. Radioparm. 33:11*, 1053–1063.
2. F. Talmont, S. Sidobre, P. Demange, A. Milon, L. Emorine. (1996) *FEBS Letts. 394*, 268–272.

NMR STRUCTURAL AND DYNAMIC STUDIES OF A DENATURED TENASCIN FIBRONECTIN TYPE III DOMAIN

Alison Meekhof, Stefan M.V. Freund and Alan Fersht

Chemical Laboratory, Gonville and Caius College, Cambridge University, Lensfield Road, Cambridge CB2 1EW, United Kingdom

The third fibronectin type III domain from human tenascin ("tenascin") is a member of the immunoglobulin superfamily, one of the largest known families of homologous structures. It is a compact all-β protein of 90 residues comprised of two adjacent β-sheets of three and four strands, respectively. We have attempted to locate regions of residual structure in denatured tenascin in order to pinpoint potential initiation sites for folding. Triple resonance experiments were used to fully assign the denatured state of ^{13}C, ^{15}N-labeled tenascin in 50 mM acetate buffer, pH 5.0, and 5 M urea at 303 K. Chemical shift deviations from random coil values were calculated for the N, NH, H$_\alpha$, and (C$_\alpha$-C$_\beta$) resonances, and peak intensity and chemical shift variations for each residue versus temperature were monitored in a series of HSQC spectra (278–303 K). Sequential NH-NH NOE intensities were measured at 278 and 303 K through HSQC-NOESY-HSQC experiments. In addition, ^{15}N R1 and R2 rates, and {^{1}H-^{15}N} NOEs were measured at 278 and 303 K. A reduced spectral density mapping approach was used to extract dynamic information from these data. Together, the results suggest three major areas of residual structure. In conjunction with protein engineering strategies, these methods will offer insight into folding mechanisms of all-β proteins.

TOWARDS A STRUCTURAL UNDERSTANDING OF GENETIC DISEASES: THE SOLUTION STRUCTURE OF THE KH DOMAIN OF FMR1

Giovanna Musco[1], Hakim Kharrat[1], Gunter Stier[1], Franca Fraternali[1], Michael Nilges[1], Toby J. Gibson[2] and Annalisa Pastore[2]

[1]EMBL, Meyerhofstr. 1 - D-69012 Heidelberg, Germany and [2]Universite de Rouen, UFR des Sciences et des Techniques Laboratoire de RMN, 76821 Mont-Saint-Aignan Cedex, France

Conventional methods to solve 3-D structures by NMR techniques rely on identification of spin systems, sequence specific assignments of the resonance in the proton spectra, compilation of intramolecular distances from two-dimensional homonuclear, three and four dimensional heteronuclear edited NOESY spectra, structures calculations by distance geometry (DG) and/or simulated annealing techniques (SA) [1]. In the last few years, automated sequence specific assignment procedures have been suggested [2–6], but they remain, at the best, only semi-automatic and a detailed inspection of the spectra by the researcher is always necessary. New approaches have been proposed to overcome this problem by semiautomated and fully automated assignment procedures for unambiguous and ambiguous NOE data [5,7]. A novel approach, the ARIA protocol, based on the XPLOR package is applied here for structure determination. This method progressively includes, by an iterative procedure, distance restraints directly assigned on the basis of chemical shift and NOE intensity lists.

In the present work we apply this new approach to the resolution of the structure of the first KH motif of FMR1, the protein involved in the fragile X syndrome. The fragile X syndrome is an X chromosome-linked dominant disorder and the most frequent heritable cause of mental retardation in humans, with an incidence of approximately 1 in 1200 males and 1 in 2500 females. Appearance of the syndrome correlates with the presence of a large trinucleotide expansion (CGG) and with the methylation of a CpG island within the promoter region of the FMR1 gene [8]. Sequence analysis of the FMR1 protein has suggested that RNA-binding might be related to the presence of two KH modules. The K-homology (KH) module is an evolutionarily conserved sequence motif of about 70 aminoacids [9]. It was originally identified in single or multiple copies in a wide variety of quite diverse proteins which all work in close association with RNA. Accumulating evidences suggest that the KH module itself is an RNA binding motif. The structure of a representative KH motif from vigilin has been recently solved [10]. This work allowed to redefine the boundaries of the entire KH sequence family and suggested the exact domain boundaries necessary to obtain the two KH modules of FMR1 as folded chains. The first FMR1 KH domain has been produced by recombinant techniques and its NMR spectrum fully assigned. The next step torward a structural determination of this domain is the conversion of the NMR distance restraints into a three-dimensional structure. This is the aim of the present project. Determination of the structures of the FMR1 KH domains represents the starting point of their structure/functional characterization in the complex wih RNA.

1. Nilges (1996) *Curr. Op. in Struct. Biol. 6*, 617–623.
2. Hare et al. (1994) *J. Biomol. NMR. 4*, 35–46.
3. Morelle et al. (1995) *J. Biomol. NMR. 5*, 154–160.
4. Meadows et al. (1994) *J. Biomol. NMR. 4*, 79–96.
5. Nilges et al. (1995) *J. Mol. Biol. 245*, 645–660.
6. Mumenthaler et al. (1995) *J. Mol. Biol. 54*, 465–480.
7. Nilges (1993) *Proteins 17*, 297–309.
8. Mandel (1992) *Curr. Op. in Genetics and Devel. 2*, 422–430.
9. Gibson et al. (1993) *FEBS Letts. 3*, 361–366.
10. Musco et al. (1996) *Cell 85*, 237–245.

DESIGNING *cis*-PROLINE REPLACEMENTS IN AN IMMUNOGLOBULIN VARIABLE DOMAIN

E. C. Ohage and B. Steipe

Genzentrum, Feodor-Lynen-Str. 5, 81377 Munich, Germany

The folding of a polypeptide chain under native conditions is a complex process frequently involving peptidyl-prolyl *cis-trans* isomerisation as a slow, rate limiting step. On the way from the unfolded to the native state, intermediates with incorrect Xaa-Pro peptide bond conformations may accumulate and result in aggregation or high susceptibility towards proteases. Immunoglobulin Vκ domains contain two highly conserved prolines which are known to be in *cis*-conformation in the native state. Of these, **Pro8** is in the framework and **Pro95** in the CDR3 region of the domain. Indeed, folding of the domain is significantly accelerated in the presence of the peptidyl-prolyl *cis-trans*-isomerase Cyclophilin.

We have used a homology modelling and statistical approach to design domains without *cis* peptide bonds in the native state. Guided by the consensus sequence for lambda-sub-type immunoglobulin variable domains, we have deleted Ser7 by site-directed mutagenesis. Pro95 was deleted based on a model corresponding to a rare variant of the CD3 loop. Both mutants were expressed in *E. coli* and folding kinetics and stability was measured. As predicted, and somewhat contrary to intuition, deleting Ser7 in the framework has negligible effect on the stability of the domain but results in faster refolding. Pro95 in hypervariable loop 3 plays an important role in domain architecture and cannot be deleted without significant loss in stability. Even though the double mutant is somewhat destabilized, it folds faster than the wild-type and its folding cannot be significantly accelerated by Cyclophilin anymore. These results show that a rationally designed replacement of cis-prolines is feasible and that such mutations can improve antibody folding.

MEANS FOR THE ASSIGNMENT OF CROWDED SPECTRA

Kimmo Pääkkönen

VTT Chemical Technology, P.O. Box 1401, FIN-02044 VTT, Finland

Advent of triple-resonance NMR experiments for doubly labeled proteins diminished interest for further development of methods for the assignment of homonuclear two-dimensional spectra. However, at times it is not possible to obtain a labeled sample and consequently one has to resort to the homonuclear methods.

We have worked out the sequence specific assignments for a 10 kDa protein domain. The overlap problem was tackled by acquiring cosy, rcosy, tocsys, dq and noesy spectra at 30, 40 and 50°C. Following systematically the shifts as a function of temperature allowed us to resolve most ambiguities involving amides. In this way the sequence specific assignment was obtained except for a few amino acids. The $C_\alpha H$ shifts were qualitatively classified according to the chemical shift index (CSI) to distinguish secondary structure elements. The subsequent assignment of noe spectra was carried out iteratively. An initial family of structures were computed based on a sparse set of restraints extracted from cross peaks without overlap or ambiguities. Further restraints were included when contributions to a cross peak from degenerate protons could be excluded according to the preliminary family of structures. The correctness of assignments were verified by spectra recorded from a ^{15}N-labeled sample obtained later.

EVIDENCE FROM ^{13}C1' RELAXATION RATE MEASUREMENTS OF INTERNAL MOTIONS IN THE *lac* OPERATOR FREE AND COMPLEXED

Françoise Paquet and Gérard Lancelot

Centre de Biophysique Moléculaire, CNRS, Rue Charles Sadron, 45071 Orléans Cedex 02, France

In order to study some internal dynamic processes of the *lac* operator sequence, the ^{13}C-labeled duplex:

5'd($C_0G_1C_2T_3C_4A_5C_6A_7A_8T_9T_{10}$) . d($A_{10}A_9T_8T_7G_6T_5G_4A_3G_2C_1G_0$) was used. The spreading of both the H1' and C1' resonances brought about an excellent dispersion of the ^{1}H1'-^{13}C1' correlations. The spin-lattice relaxation parameters R(Cz), R(Cx,y) and R(Hz~Cz) were measured for each residue of the two complementary strands, except for the 3'-terminal residues which were not labeled. Variation of the relaxation rates was found along the sequence. All the C1'-H1' fragments exhibited fast (J_i = 10 to 80 ps) restricted libration motions (S2 = 0.50 to 0.95). Moreover, the fit of both R(Cz) and R(Hz~Cz) experimental relaxation rates using an only global correlation time for all the residues, gave evidence of a supplementary relaxation pathway affecting R(Cx,y) for the purine residues in the (5'~3') G_4 A_3 and A_{10} A_9 T_8 T_7 sequences. This relaxation process was analyzed in terms of exchange stemming from motions of the sugar around the glycosidic bond on the millisecond time scale.

Similarly relaxation rates R(Cx,y) were measured on the *lac* operator complexed with *lac* repressor headpiece. Exchange processes were showed for T_3, A_8, A_9T_8, C_5 and G_4A_3. It should be pointed out that these nucleotide residues have already given evidence of close contacts with the protein in the complex.

SOLUTION STRUCTURE OF THE UNIVERSALLY CONSERVED 2250 LOOP OF 23S RIBOSOMAL RNA

Elisabetta Viani Puglisi, Harry F. Noller, and Joseph D. Puglisi

Center for Molecular Biology of RNA, University of California, Santa Cruz, CA 95064

Ribosomal RNA mediates the essential interaction between transfer RNA and the ribosome during translation. Catalysis of peptide bond formation occurs on the 50S ribosomal subunit, where the universally conserved 3'CCA ends of two tRNAs in the P and A sites are specifically positioned. A universally conserved hairpin loop within 23S ribosomal RNA, the 2250 loop, is required for tRNA binding in the P site. G2252 in this loop forms a Watson-Crick base pair with C74 of the tRNA 3'-end. To understand the structural basis of this essential interaction, we have determined the solution structure of the 2250 loop RNA using nuclear magnetic resonance spectroscopy (NMR). The structure reveals how a guanine-rich loop presents nucleotides for RNA-RNA interactions

MUTANT AND WILD-TYPE MYOGLOBIN-CO PROTEIN DYNAMICS: VIBRATIONAL ECHO EXPERIMENTS

K. D. Rector[1], C. W. Rella[2], Jeffrey R. Hill[3], A. S. Kwok[1], Stephen G. Sligar[3], Ellen Y. T. Chien[3], Dana D. Dlott[3] and M. D. Fayer[1]

[1]Department of Chemistry and [2]Stanford Free Electron Laser Center, Stanford University, Stanford, CA 94305; [3]School of Chemical Sciences, University of Illinois Urbana-Champaign, Urbana, IL 61801

Picosecond infrared vibrational echo experiments on a mutant protein, H64V myoglobin-CO, are described and compared to experiments on wild type myoglobin-CO. H64V is myoglobin with the distal histidine replaced by a valine. The vibrational dephasing experiments examine the influence of protein dynamics on the CO ligand, which is bound to the active site of the mutant protein, from low temperature to physiologically relevant temperatures. The experiments were performed with a mid-infrared free electron laser tuned to the CO stretch mode at 1969 cm^{-1}.

The vibrational echo results are combined with infrared pump-probe measurements of the CO vibrational lifetime to yield the homogeneous pure dephasing. The homogeneous pure dephasing is the Fourier transform of the homogeneous linewidth with the lifetime contribution removed. The measurements were made from 60 K to 300 K and show that the CO vibrational spectrum is inhomogeneously broadened at all temperatures studied. The mutant protein's CO vibrational pure dephasing rate is ~20% slower (narrower homogeneous pure dephasing linewidth) than the wild type protein at all temperatures, although the only difference between the two proteins is the replacement of the wild type's polar distal histidine amino acid by a non-polar valine. These results provide insights into the mechanisms of the transmission of protein fluctuations to the CO ligand bound at the active site, and they are consistent with previously proposed mechanisms of protein-ligand coupling.

^{1}H-NMR RESONANCES ASSIGNMENT AND SECONDARY STRUCTURE OF BINASE

M. J. Reibarkh, D. E. Nolde, E. V. Bocharov, L. I. Vasilieva and A. S. Arseniev

Shemyakin and Ovchinnikov Institute of Bioorganic Chemistry, Russian Academy of Sciences ul. Miklukho-Maklaya 16/10, 117871, Moscow, Russia

Near complete resonance assignment was obtained in two-dimensional ^{1}H-NMR spectra of guanilspecific ribonuclease from *Bacillus Intermedius* (binase). The hydrogen-deuterium exchange rates of amide protons were measured in ^2H_2O at pH 6.7 and 30°C. The secondary structure of binase was determined based on vicinal coupling constants $^3J_{HNC\alpha H}$, NOE contacts, amide protons exchange rates, and chemical shift indexes of $C^\alpha H$-protons. The binase consists of 3 α-helices in N-terminal part of the molecule, including residues 6–16, 26–31, 41–45, and a 5-stranded β-sheet, including residues 51–55, 71–75, 86–90, 95–99, and 104–108. The secondary structure of binase contains a β-bulge formed by Gly52 and Asp53. The secondary structures of binase and its nearest homologue, barnase from *Bacillus Amiloliquefaciens*, are compared.

AN *ab initio* APPROACH TO PREDICT AN ACTIVE SITE: A CASE STUDY OF ASPARAGINE SYNTHETASE

Sangeeta Sawant,[1] Sheldon M. Schuster,[2] and A.S. Kolaskar[1]

[1]Bioinformatics Centre, University of Pune, Ganeshkhind, Pune, India and
[2]Interdisciplinary Centre for Biotechnology Research, University of Florida, Gainesville, FL 32610

An approach based on following four main steps has been developed to identify active site of an enzyme. These are:

1. Locating the invariant regions in a family of proteins from different biological sources;
2. predicting the three dimensional structure of such region using Molecular modelling approach;
3. docking of substrate/competitive inhibitor onto this modelled molecule and
4. confirmation of the active site through site-directed mutagenesis studies in laboratory. This method was applied to predict the aspartate binding site of the enzyme asparagine synthetase.

Asparagine synthetase is a key enzyme which increases the levels of asparagine in leukamic patients and thus a specific inhibitor of this enzyme can help in treatment of such patients. Few physico-chemical or kinetic studies are available for this enzyme. Sequences of Asparagine synthetase from 6 different sources are aligned and invariant regions are identified. A threonine-rich invariant region around amino acids 316–332 is postulated to be the aspartate-binding site. Therefore molecular modelling and simulation of a 12-mer peptide soaked in water were carried out using the Biosym/MSI software InsightII and Discover. For the simulations, a simulated annealing approach was used. The global minimum free energy conformation was found to be folded; having a cavity in which the aspartate molecule was docked. The docking interaction energy was calculated.

In the laboratory, several single point and double mutations were carried out in this region of the protein. It was found that the binding of cofactors glutamine and ATP does not get affected due to any mutations, though the aspartate binding is drastically affected in some of the mutated peptides. This indicates that the local conformation is altered due to mutations which was also observed in the molecular modelling studies of the 12mer peptide.

This approach, we feel is a general one and can be used in identifying active sites/binding sites on enzymes/proteins. Details of the approach and results will be discussed.

THERMAL UNFOLDING STUDIES OF TWO DIMERIC TRANSCRIPTIONAL FACTORS TEF (THYROTROPH EMBRYONIC FACTOR) AND DBP (ALBUMIN SITE-D BINDING PROTEIN)

Olivier Schaad[1], Philippe Fonjallaz,[2] Marius Clore,[3] Angela Gronenborn,[3] and Ueli Schibler[2]

[1]Depts. of Biochemistry and [2]Molecular Biology, SCIENCES - II, University of Geneva, 1211 Geneva 4, Switzerland and [3]Lab. of Chemical Physics NIH/NIDDK Bethesda, MD

Thermal unfolding studies of two dimeric transcriptional factors TEF (Thyrotroph Embryonic Factor) and DBP (Albumin site - D Binding Protein) were performed to evaluate their dimerization constants, KD. Circular dichroïsm measurements showed that both proteins unfold with a cooperative transition around 50°C if they have no SS bridge between dimers, but unfold only partially in the temperature range 15°C-60°C if the cysteines are oxidized. We use three different models in order to evaluate KD:

1. D Fully Folded $\rightarrow$ M Fully Unfolded
2. D Fully Folded $\rightarrow$ D partial Folded (Zipper Only) $\rightarrow$ M Fully Unfolded
3. D Fully Folded $\rightarrow$ M Folded $\rightarrow$ M Fully Unfolded

Our results suggest that Model 2 provides the most satisfactory representation of the data. It involves a basic helical domain that is not very stable without DNA (with a TM around 18°C) and a more stable region the leucine zipper region (with a TM around 50°C). This model is compatible with the scissor grip type of binding reported for GCN4 and postulated for other leucine zipper DNA binding proteins such as fos.

HIERARCHICAL PROTEIN STRUCTURE SUPERPOSITION

Amit P. Singh and Douglas L. Brutlag

Section on Medical Informatics and Department of Biochemistry, Stanford University, Stanford, CA 94305

The structural comparison of proteins has become increasingly important as a means to identify protein motifs, classify proteins into fold families, and predict protein function. We have developed a new algorithm for the superposition of protein structures based on a hierarchical decomposition of the structures at the secondary structure and atomic level. Our technique represents secondary structure elements as vectors and defines a set of orientation independent and orientation dependent scoring functions to compare vectors from the two proteins. These scores are then used in a dynamic programming algorithm that computes an optimal local alignment of the secondary structure vectors from the target and query proteins. The secondary structure superposition is followed by an atomic superposition stage which iteratively minimizes the RMSD between pairs of nearest atoms from the two proteins. The final stage of our algorithm identifies a core of well aligned atoms and refines the previous atomic alignment by minimizing the RMSD of these core atoms. This hierarchical technique of first searching among secondary structure alignments and then among atomic level alignments gives our technique the increased flexibility of detecting global as well as local similarities. We compared our method to Holm and Sander's DALI algorithm and found that we were able to detect structural similarity at the same level as DALI. We also searched a representative set of the Protein Data Bank using various proteins and protein motifs as queries and detected structural similarity between several distantly related proteins.

THREE DIMENSIONAL STRUCTURE IN SOLUTION OF *Ace*-AMP1, A POTENT ANTIMICROBIAL PROTEIN EXTRACTED FROM ONION SEEDS: STRUCTURAL ANALOGIES WITH PLANT NONSPECIFIC LIPID TRANSFER PROTEINS

S. Tassin, P. Sodano, W. F. Broekaert, D. Marion, F. Vovelle, and M. Ptak

Centre de Biophysique Moleculaire, CNRS, 2 rue Charles Sadron, 45071 Orléans Cédex 02, France

Since the initial characterization of thionins in 1972, other antibiotic proteins have been isolated from plants such as β-1,3-glucanases, permatins, ribosome-inactivating proteins, cysteine-rich antimicrobial peptides, plant defensins, albumins and several non-specific lipid transfer proteins. *Ace*-AMP1 was isolated from onion seeds and display a potent activity against numerus fungi and bacteria. This cationic protein contains 93 amino-acid residues and four disulfide bridges. This protein is homologous to plant non-specific lipid transfer proteins (ns-LTP) which transfer lipids between two membranes *in vitro*. The structures of several ns-LTP have been determined by NMR or X-ray diffraction. Its main structural feature is the presence of an internal cavity running through the whole molecule which is the binding site of lipids. *Ace*-AMP1 was studied in solution by NMR and its structure calculated from 1300 distance constraints. The global fold of *Ace*-AMP1 is similar to the ns-LTPs fold, involving four helices wound on a right handed super-helix. The most striking feature of *Ace*-AMP1 is the absence of any continuous cavity running through the molecule as described in ns-LTP structures. Two aromatic residues belonging to H4 helix and the C-terminal region are oriented towards the interior of the protein in the hydrophobic channel and obstruct the cavity. The absence of a continuous hydrophobic cavity may explain why this protein is unable to transfer lipids.

STRUCTURE OF LEGHEMOGLOBIN MOLECULE AND ITS FUNCTION AS AN OXYGEN-CARRYING PROTEIN

A. F. Topunov and N. E. Petrova

A. N. Bach Institute of Biochemistry, Russian Academy of Sciences, Leninsky prospect 33, Moscow 117071, Russia

Leghemoglobin (Lb) - plant hemoglobin is an oxygen-carrier in nitrogen-fixing legume nodules. Thanks to its structure it can function in specific conditions of nodules (very low oxygen concentration). Lb is a monomeric protein containing a standard heme. It has very high affinity to oxygen, probably because of the structure of its heme pocket. Lb can also bind other ligands and some of them can be of physiological significance, for example NO, after reaction with which nitrozo-Lb appears. It is interesting that some leghemoglobins (e.g. lupin Lb) present the Bohr effect on the level of monomeric structure.

We proposed a scheme of Lb functioning which describes the interaction of different processes with its participation. In this scheme we take into account main processes, such as Lb reduction, oxidation, oxygenation and deoxygenation and some minor ones (for instance production of the so-called ferryl Lb form). Possible ways of processes named were also examined. We especially studied the process of enzymatic reduction of Lb and the ezyme catalyzing it. For description of details of Lb functioning and their distinctions from other hemoglobins structural data of Lb molecule were used.

PREPARATION OF L-ASPARAGINE COMPLEX OF TECHNETIUM-99m

Perihan Unak and Fatma Yurt

Ege University, Institute of Nuclear Sciences, Bornova - IZMIR, 35100 Turkey

Various methods were experimented with to obtain the complex of L-asparagine with 99mTc. First the VII state of Tc was reduced with direct reduction by SnCl2 and the complex compound couldn't be obtained. For this reason, ligand exchange with 99mTc-gluconate complex was tried and a 20% yield was obtained. 99mTc-pyrophosphate complex was carried out as a second ligand exchanger and a 50% yield was obtained in this experiment. Optimum conditions for ligand exchange with pyrophosphate are 11 for pH, 50°C for temperature, 60 minutes for reaction time and 278/1 for pyrophosphate concentration to SnCl2 ratio.

COMPUTER MODELING OF ACETYLCHOLINESTERASE ACTIVITY

Stanislaw T. Wlodek,[1] Terry W. Clark,[1] L. Ridgway Scott,[1] J. Andrew McCammon,[2] James M. Briggs,[2] and Jan Antosiewicz[3]

[1]Texas Center for Advanced Molecular Computation, University of Houston, TX 77204-5502 and [2]Department of Chemistry and Biochemistry, University of California at San Diego, La Jolla, CA 92092-0365; [3]Department of Biophysics, University of Warsaw, Warsaw 02-089, Poland

The dynamic properties of acetylcholinesterase dimer from *Torpedo californica* liganded with tacrine (TcAChE-THA) have been studied using molecular dynamics (MD). The simulation reveals fluctuations in the width of the primary channel to the active site that are large enough to admit substrates. Alternative entries to the active site through the side walls of the gorge have been detected in a number of structures, suggesting a mechanism by which the enzyme could remain active with the primary entrance blocked. A large scale motion of slight contraction and relative rotation of the protein subunits has been detected.

In a separate study, Brownian dynamics simulations of the encounter kinetics between the active site of wild-type and the Glu199 mutant of TcAChE with a charged substrate were performed. In addition, *ab initio* quantum chemical calculations were undertaken of the energetics of transformation of the Michaelis complex into a covalently bound tetrahedral intermediate in models of the wild-type and Glu199Gln mutant active sites. The calculations predicted about a factor of 30 reduction in the rate of formation of the tetrahedral intermediate upon the Glu199Gln mutation, and showed that the Glu199 residue located in the proximity of the enzyme active triad boosts AChE's activity in a dual fashion: (1) by increasing the encounter rate due to the favorable modification of the electric field inside the enzyme reaction gorge and (2) by stabilization of the transition state for the first chemical step of catalysis. Our calculations also demonstrate the critical role of the oxyanion hole in stabilization of the tetrahedral intermediate.

THREE DIMENSIONAL NMR STUDIES OF DNA RECOGNITION BY REGULATORY FACTOR II

Yate-Ching Yuan[1], Robert H. Whitson, Jr.[2], Keiichi Itakura[2] and Yuan Chen[1]

[1]Division of Immunology and [2]Division of Biology, Beckman Research Institute of the City of Hope, Duarte, CA 91010, USA

The modulator regulatory factor II (MRF-2) is a sequence specific DNA binding protein. It recognizes the transcription modulator of the major immediate-early gene of human cytomegalovirus (HCMV). It belongs to a new family of the DNA binding proteins. The DNA binding domain of this protein, free and in complex with DNA, is being studied by three dimensional heteronuclear NMR methods. The DNA binding domain is uniformly labeled with ^{15}N and ^{13}C and resonance assignments have been obtained by the multidimensional triple resonance approach. Conformational constraints are obtained from ^{13}C- and ^{15}N-edited NOESY spectra and coupling constant measurements. The solution structure of the DNA binding domain of Mrf-2 is not homologous to the known DNA binding motifs. High-resolution solution structures and molecular dynamics of the protein will be discussed.

STRUCTURAL FEATURES OF β-HAIRPIN PEPTIDE

Rosa Zerella, Benjamin Beardsley, Philip A. Evans, and Dudley H. Williams

Department of Chemistry, St. John's College, University of Cambridge, Cambridge CB2 1TP, UK

Short peptides are often used to model folding pathways. Measuring the folding states of isolated peptides can suggest the role of these structural elements in the folding pathway of the full length protein. The conformation of small peptides is often described as a two state system: one state is completely folded and native-like, and the other is random coil. An alternative characterization is to regard the peptide structure as flexible within an envelop surrounding the native structure. Careful analysis of NMR data acquired on a 17 amino acid peptide derived from the N-terminus of ubiquitin supports the second description of peptide structure. A native-like β-hairpin structure is formed by this peptide in aqueous solution. NMR observables such as chemical shift, coupling constants and NOEs have been measured and evaluated for their concordance with either the two state hypothesis or the multi state hypothesis. Taken together, the data support the concept that this short peptide is fully structured yet flexible. This conclusion requires the re-evaluation of other studies concerning peptide models for protein folding: often, the fraction of folded peptide has been used to infer the importance of a particular structural element in protein folding, but this study suggests that these fractions are likely to be wrong if calculated from only one NMR parameter. Studies on a series of glycopeptide antibiotics support this conclusion.

INDEX